Photon-Hadron Interactions

Advanced Book Classics

Anderson: Basic Notions of Condensed Matter Physics, ABC ppbk, ISBN 0-201-32830-5

Atiyah: K-Theory, ABC ppbk, ISBN 0-201-40792-2

Bethe: Intermediate Quantum Mechanics, ABC ppbk, ISBN 0-201-32831-3

Clemmow: Electrodynamics of Particles and Plasmas, ABC ppbk, ISBN 0-20147986-9

Davidson: Physics of Nonneutral Plasmas, ABC ppbk ISBN 0-201-57830-1

DeGennes: Superconductivity of Metals and Alloys, ABC ppbk, ISBN 0-7382-0101-4

d'Espagnat: Conceptual Foundations Quantum Mechanics, ABC ppbk, ISBN 0-7382-0104-9

Feynman: Photon-Hadron Interactions, ABC ppbk, ISBN 0-201-36074-8

Feynman: Quantum Electrodynamics, ABC ppbk, ISBN 0-201-36075-4

Feynman: Statistical Mechanics, ABC ppbk, ISBN 0-201-36076-4

Feynman: Theory of Fundamental Processes, ABC ppbk, ISBN 0-201-36077-2

Forster: Hydrodynamic Fluctuations, Broken Symmetry, and Correlation Functions, ABC ppbk, ISBN 0-201-41049-4

Gell-Mann/Ne'eman: The Eightfold Way, ABC ppbk, ISBN 0-7382-0299-1

Gottfried: Quantum Mechanics, ABC ppbk, ISBN 0-201-40633-0

Kadanoff/Baym: Quantum Statistical Mechanics, ABC ppbk, ISBN 0-201-41046-X

Khalatnikov: An Intro to the Theory of Superfluidity, ABC ppbk, ISBN 0-7382-0300-9

Ma: Modern Theory of Critical Phenomena, ABC ppbk, ISBN 0-7382-0301-7

Migdal: Qualitative Methods in Quantum Theory, ABC ppbk, ISBN 0-7382-0302-5

Negele/Orland: Quantum Many-Particle Systems, ABC ppbk, ISBN 0-7382-0052-2

Nozieres/Pines: Theory of Quantum Liquids, ABC ppbk, ISBN 0-7382-0229-0

Nozieres: Theory of Interacting Fermi Systems, ABC ppbk, ISBN 0-201-32824-0

Parisi: Statistical Field Theory, ABC ppbk, ISBN 0-7382-0051-4

Pines: Elementary Excitations in Solids, ABC ppbk, ISBN 0-7382-0115-4

Pines: The Many-Body Problem, ABC ppbk, ISBN 0-201-32834-8

Quigg: Gauge Theories of the Strong, Weak, and Electromagnetic Interactions, ABC ppbk, ISBN 0-201-32832-1

Richardson: Experimental Techniques in Condensed Matter Physics at Low Temperatures, ABC ppbk ISBN 0-201-36078-0

Rohrlich: Classical Charges Particles, ABC ppbk ISBN 0-201-48300-9

Schrieffer: Theory of Superconductivity, ABC ppbk ISBN 0-7382-0120-0

Schwinger: Particles, Sources, and Fields Vol. 1, ABC ppbk ISBN 0-7382-0053-0

Schwinger: Particles, Sources, and Fields Vol. 2, ABC ppbk ISBN 0-7382-0054-9

Schwinger: Particles, Sources, and Fields Vol. 3, ABC ppbk ISBN 0-7382-0055-7

Schwinger: Quantum Kinematics and Dynamics, ABC ppbk, ISBN 0-7382-0303-3

Thom: Structural Stability and Morphogenesis, ABC ppbk, ISBN 0-201-40685-3

Wyld: Mathematical Methods for Physics, ABC ppbk, ISBN 0-7382-0125-1

Photon-Hadron Interactions

Richard P. Feynman
late, California Institute of Technology

CRC Press
Taylor & Francis Group
Boca Raton London New York

CRC Press is an imprint of the
Taylor & Francis Group, an **informa** business

Advanced Book Program

First published 1972 by Westview Press

Published 2018 by CRC Press
Taylor & Francis Group
6000 Broken Sound Parkway NW, Suite 300
Boca Raton, FL 33487-2742

Visit the Taylor & Francis Web site at
http://www.taylorandfrancis.com

and the CRC Press Web site at
http://www.crcpress.com

Library of Congress Cataloging-in-Publication Data

Feynman, Richard Phillips.
Photon-hadron interactions / Richard Feynman.
p. cm. — (Advanced book classics series)
Originally published : Reading, Mass. : W. A. Benjamin, Advanced Book Program, 1972. (Frontiers in physics)
Includes index.
1. Photon-hadron interactions. 2. Hadron interactions.
I. Title. II. Series.
QC793.5.P428F49 1989 539.7'54—dc19 89-30697

ISBN 13:978-0-201-36074-5 (pbk)

Cover design by Suzanne Heiser

Editor's Foreword

Addison-Wesley's *Frontiers in Physics* series has, since 1961, made it possible for leading physicists to communicate in coherent fashion their views of recent developments in the most exciting and active fields of physics—without having to devote the time and energy required to prepare a formal review or monograph. Indeed, throughout its nearly forty-year existence, the series has emphasized informality in both style and content, as well as pedagogical clarity. Over time, it was expected that these informal accounts would be replaced by more formal counterparts—textbooks or monographs—as the cutting-edge topics they treated gradually became integrated into the body of physics knowledge and reader interest dwindled. However, this has not proven to be the case for a number of the volumes in the series: Many works have remained in print on an on-demand basis, while others have such intrinsic value that the physics community has urged us to extend their life span.

The *Advanced Book Classics* series has been designed to meet this demand. It will keep in print those volumes in *Frontiers in Physics* or its sister series, *Lecture Notes and Supplements in Physics*, that continue to provide a unique account of a topic of lasting interest. And through a sizable printing, these classics will be made available at a comparatively modest cost to the reader.

These notes on Richard Feynman's lectures at Caltech on the topic of photon-hadron interactions, in which he developed his theory of partons, were first published some twenty-five years ago. As is the case with all of Feynman's lectures, the presentation in this work reflects his deep physical insight, the freshness and originality of his approach to understanding physics, and his overall pedagogical wizardry. As a result, this volume will always be of fundamental importance to anyone interested in understanding the development of quantum chromodynamics (QCD)—the theory of quarks and gluons—which explains hadron-hadron interactions at high energies.

David Pines
Urbana, Illinois
December 1997

Special Preface

Many of us were first introduced to the concepts of the parton model from this book. At that time the competing view was one in which there were no elementary particles. Every particle was supposed to be a composite of every other particle. The ideas and concepts in this book have helped pave the way for our understanding of the constituent nature of hadrons which eventually led to the Quantum Chromodynamic (QCD) theory of quarks and gluons. As is true of most of Feynman's books, the maximum benefit is obtained if one has previously studied the subject in some detail. Feynman's unique perspective can best be appreciated by readers with a solid background in the subject.

Although this book is almost 18 years old, it still is an excellent reference. It appears on the recommended reading list of all the current QCD books. The book provides a good understanding of the parton model from the man who invented it. In the "pre-QCD" or "naive" parton model the constituents within hadrons were assumed to be bounded in the transverse direction. The probability of finding a parton within a high momentum hadron with a large transverse momentum was assumed to fall like a Gaussian or an exponential. QCD tells us that this is not exactly true and gives a power law fall-off in the transverse momentum. Because of this, many "naive" parton model expectations are modified (in an *important* way) by logarithmic factors. Feynman used to laugh when his parton model was referred to as "naive," and he would say, "At least I got it right up to logarithms." We all miss Feynman very much and it is through books like this that his ideas live on.

R.D. Field
March, 1989

Vita

Richard P. Feynman

Born in 1918 in Brooklyn, Richard P. Feynman received his Ph.D. from Princeton in 1942. Despite his youth, he played an important part in the Manhattan Project at Los Alamos during World War II. Subsequently, he taught at Cornell and at the California Institute of Technology. In 1965 he received the Nobel Prize in Physics, along with Sin-Itero Tomanaga and Julian Schwinger, for his work in quantum electrodynamics.

Dr. Feynman won his Nobel Prize for successfully resolving problems with the theory of quantum electrodynamics. He also created a mathematical theory which accounts for the phenomenon of superfluidity in liquid helium. Thereafter, with Murray Gell-Mann, he did fundamental work in the area of weak interactions such as beta decay. In later years Feynman played a key role in the development of quark theory by putting forward his parton model of high energy proton collision processes.

Beyond these achievements, Dr. Feynman introduced basic new computational techniques and notations into physics, above all, the ubiquitous Feynman diagrams which, perhaps more than any other formalism in recent scientific history, have changed the way in which basic physical processes are conceptualized and calculated.

Feynman was a remarkably effective educator. Of all his numerous awards, he was especially proud of the Oersted Medal for Teaching which he won in 1972. *The Feynman Lectures on Physics*, originally published in 1963, were described by a reviewer in Scientific American as "tough, but nourishing and full of flavor. After 25 years it is *the* guide for teachers and for the best of beginning students." In order to increase the understanding of physics among the lay public, Dr. Feynman wrote *The Character of Physical Law* and *Q.E.D.: The Strange Theory of Light and Matter*. He also authored a number of advanced publications that have become classic references and textbooks for researchers and students.

Richard Feynman died on February 15, 1988.

Contents

Preface

The most advanced course in graduate theoretical physics at Caltech is "Special Topics in Theoretical Physics." Each year the professor chooses the topic with which he will deal. This year (1971-72), having just come back from the 1971 International Symposium on Electron and Photon Interactions at High Energies, held at Cornell University, my own interest in the subject was aroused, and I chose to analyze the various theoretical questions related to that conference. The lectures themselves became so extensive that the decision was made to put them into book form, with the thought that other people might also be interested. Thus, the report of the Cornell conference should be considered as a companion volume to these lecture notes. The references given here are far from complete, but a full list of references is given in the Proceedings of the Symposium, published by the Laboratory of Nuclear Studies, Cornell University, January 1972.

The material is dealt with on an advanced level; for instance, knowledge of the theory of hadron-hadron interactions is assumed. I have tried to analyze in detail where we stand theoretically today. The treatment is somewhat uneven; for example, I should have liked to study the theory of the decay of the   in more detail than I was able to do. On the other hand, there are long discussions of vector meson dominance and of deep inelastic scattering. The possible consequences of the parton model are fully discussed.

Time did not permit me to complete the original plan which was to include the theory of weak interaction currents which are so closely related to electromagnetic currents.

Many thanks must go to Arturo Cisneros who edited, corrected, and extended the lectures from my class notes. Without his effort, this book would not have been possible. I also wish to thank Mrs. Helen Tuck for typing the lecture notes.

Richard P. Feynman
Pasadena, California
Summer 1972

Photon-Hadron Interactions

General Theoretical Background

Lecture 1

One very powerful way of experimentally investigating the strongly interacting particles, (hadrons) is to look at them, to probe them with a known particle; in particular the photon (no other is known as well). This permits a much finer control of variables, and probably decreases the theoretical complexity of the interactions. For example in an ordinary hadron-hadron collision like $\pi p \to \pi p$ we are hitting two unknowns together, and further, we can only vary the energy, we cannot vary the q^2 of the pion which must be m_π^2. In fact a "pion far off its mass shell" may be a meaningless - or at least highly complicated idea. On the other hand in $\gamma + p \to p + \pi$ we know the γ is single and definite, and we can vary the q^2 of the γ by using virtual γ's via, for example, electron scattering $e + p \to e + p + \pi$.

We are assuming that we do know the photon. QED has been checked so closely that we know that if the photon propagator were off by a factor of the form $(1 - q^2/\Lambda^2)^{-1}$ then Λ exceeds 4 or 5 GeV. The amplitudes are known to about 5% for q^2 as high as $(1\ \mathrm{GeV})^2$. For the rest of this course we shall assume QED is exact. There is already evidence, as we shall see, that in virtual photon-hadron collisions the photon acts normally (i.e., obeys QED expectations) up to Λ of 6 to 8 GeV.

At any rate we shall suppose QED exact - where we mean by QED the standard interaction theory for electrons, muons and photons. Exact, but incomplete - for hadrons are charged and interact also with the QED system. We discuss first how we shall assume we can describe this interaction.

Since e^2 is small it is natural to describe their interaction in a series of orders in e. One photon exchange, two photon exchange, etc. (It might be thought that to describe this coupling we shall have to have some detailed dynamical theory of the hadrons - ultimately, of course, yes - but some things can be said in general restricting the matrix elements whatever the underlying hadron dynamics - and it is these restrictions we seek in this lecture.)

The no coupling case presents no problem. The factor giving the amplitude that a hadron system goes from an incoming state $|n, in\rangle$, to an outgoing state $\langle m, out|$ is:

$$S_{mn} = \langle m, out \mid n, in\rangle \tag{1.1}$$

The S matrix is the transformation matrix from the "in" representation to the "out" representation

$$\sum_n S_{mn} \langle n, in| = \langle m, out| \tag{1.2}$$

The state $|n, in\rangle$ means a state which far in the past is asymptotically free stable hadrons (stable in strong interactions only, e.g. π^o is "stable") described by momenta, and helicities all contained via the index n. The state m, out has the set of indices m with the same space of indices, but represents a state which in the future is asymptotically in situation m.

Thus the S matrix is really the unit matrix but in a mixed representation, a different labeling for incoming and outgoing states. Supposing these states are all there are, conservation of probability requires

$$S^+S = 1. \quad (\text{i.e., } \sum_m (S_{mn'})^* S_{mn} = \delta_{nn'}) \tag{1.3}$$

(In the special case that the state n represents a single stable particle the in and out states are the same).

<u>First Order Coupling.</u>

The general coupling of electrons and hadrons is represented by the diagram:

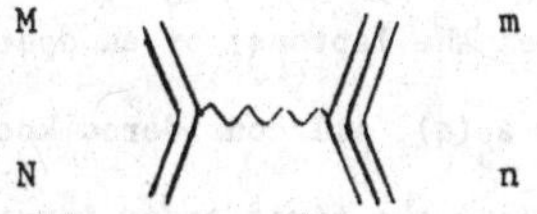

The electron-photon system goes from state N to M, the hadrons from n in to m out. We suppose the only interaction possible is by the exchange of a photon - and this photon is characterized by a polarization μ, momentum q:

$$\text{amp} = \langle M|j_\mu(q)\,|\,N\rangle \;\frac{4\pi e^2 i}{q^2}\; \langle m,\text{ out}|J_\mu(q)|n,\text{ in}\rangle\ . \tag{1.4}$$

That is to say (supposing we could measure the amplitude) we define in a given experiment the quantity

$$\mathcal{J}_\mu(q)_{mn} = \langle m,\text{ out}|J_\mu(q)|n,\text{ in}\rangle \quad . \tag{1.5}$$

This is done by removing from the measured amplitude the known (by QED theory) factors

$$\langle M|j_\mu(q)|N\rangle = 4\pi e^2 i/q^2 .$$

It is then our first supposition that this quantity $\mathcal{J}_\mu(q)_{mn}$ depends only on the states m, n of the hadron system and only the virtual momentum q and polarization μ of the virtual photon. $\mathcal{J}_\mu$ depends in no way on how the photon was made (eg. whether by μ's or electrons or on the angles and energies of the electron for fixed q and photon polarization).

This is a strong assumption. It has been verified most completely for the case of proton form factor measurements, but is often assumed in checking equipment, comparing results from one lab to another etc. We assume it.

We emphasize then that $\mathcal{J}_\mu(q)$ is an experimentally defined quantity - definable in principle for all q.

We find it convenient to define a new matrix $J_\mu(q)_{kn}$ defined in a non-mixed representation as

$$J_\mu(q)_{kn} = \langle k,\text{ in}|J_\mu(q)|n,\text{ in}\rangle \tag{1.6}$$

so we now write

$$\mathcal{J}_\mu(q)_{m,n} = \sum_m S_{mk} J_\mu(q)_{kn} \qquad (1.7)$$

To deal with this a little more abstractly, the quantity $j_\mu \cdot \frac{4\pi e^2}{q^2}$ can, in QED be described as the matrix element (between the leptons) of an operator, the vector potential operator for the leptons $a_\mu(q)$, all considered known and compatible by QED in any specific case. Thus the first order interaction is described by the matrix

$$1 + i\int a_\mu(q)\ J_\mu(q)\ d^4q + \text{higher order.}$$

The 1 comes from the zero order (where we saw the S matrix is really the unit matrix in an unmixed representation).

Unitarity requires to first order ($a^+ = a(-q)^*$ transpose)

$$(1 - i\int a_\mu^+(q)\ J_\mu^+(q)\ d^4q + \cdot\cdot)\ (1 + i\int a_\mu(q)\ J_\mu(q)\ d^4q +) = 1 \qquad (1.8)$$

or, since $a_\mu(q)$ is arbitrary, $J_\mu^+ = J_\mu$. J_μ is hermitian. Since all q are available we can define Fourier Transforms

$$J_\mu(x, y\ z\ t) = \int e^{-iq\cdot x}\ J_\mu(q)\ dq^4/(2\pi)^4$$

(and for a_μ/space).

Thus the coupling is

$$\int a_\mu(1)\ J_\mu(1)\ d\tau_1.$$

Lecture 2

Conservation of Current.

You may, if you wish simply assume $\nabla_\mu J_\mu = 0$, conservation of current for hadrons, or else note the following discussion.

It is not, strictly, true that $\mathcal{J}_\mu$ from Eq. (1.5) can be completely obtained from experiment. That is because $a_\mu(q)$ is not completely arbitrary. When it comes from the usual diagram rules it always satisfies $q_\mu a_\mu(q) = 0$. Thus one component of a_μ (the one in direction q_μ) is always missing (unless

$q^2 = 0$) and thus one component of J_μ, the component in the direction q_μ is missing. We finish the definition of J_μ by choosing $q_\mu J_\mu = 0$.

This we do in the following way. First, for $q^2 = 0$, a free proton of polarization e_μ the coupling is $e_\mu J_\mu(q)$ - but for a free photon e_μ is undefined - to it could be added αq_μ ($e'_\mu = e_\mu + \alpha q_\mu$) any α. This can make no effect so our consistency with QED demands $q_\mu J_\mu = 0$ at least when $q^2 = 0$. This is a physical property J_μ must satisfy. If it is not true for general q^2 redefine a new J_μ' to replace the old via $J_\mu' = J_\mu - q_\mu(q_\nu J_\nu)/q^2$. Evidently $q_\mu J_\mu' = 0$, and no new pole at $q^2 = 0$ is introduced by the new $1/q^2$ term for its numerator $q_\nu J_\nu$ vanishes at $q^2 = 0$.

What of other restrictions on $J_\mu(1)$? These we wish to find. For example since it is a local operator field theory suggests that if hadrons are governed by an underlying field theory then $\left[J_\mu(1), J_\nu(2)\right] = 0$ if 1 and 2 are space like separated (symbolized by $\times$ 1 • 2). If you wish you may assume this - but it is very interesting that we can prove it from our assumption that the strong system interacts with QED (subject to systematic errors in proof due to tacit assumptions. This may not be an important point but it is interesting so I will waste your time by proving it).

<u>2nd Order Coupling.</u>

This is via diagrams of the type whose amplitude depends on a computable

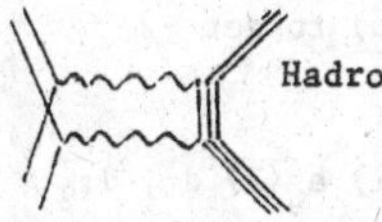

factor from the leptons* times a matrix element depending on the two momenta and polarizations of the two virtual photons $-1/2\ V_{\mu\nu}(q_1, q_2)$. As defined this is symmetrical in $q_1 \leftrightarrow q_2$, $\mu \leftrightarrow \nu$, for Bose statistics among photons does not permit us to distinguish the photons so no other function can be experimentally defined. Using an unmixed representation, and coordinate space (via double Fourier transform) we can represent this amplitude as

$$-\frac{1}{2}\int\int V_{\mu\nu}(1, 2)\left\{a_\mu(1)\ a_\nu(2)\right\}_T d\tau_2\ d\tau_1 \quad . \tag{2.1}$$

---

*In part I of this course "leptons" will mean e^-, e^+, μ^-, μ^+ only.

The contribution from the QED lepton side can be figured for each diagram, and can, as shown in QED theory, be generally written as the matrix element (between in and out photon-lepton states) of the time ordered product of the operator $a_\mu(1)$ and $a_\mu(2)$ (symbolized by $\{\}_T$). Since arbitrary $a(1)a(2)$ can be made $V_{\mu\nu}$ is experimentally defined.

I wish now to prove a number of things, so it will be more convenient to restrict the integral to $t_1 > t_2$ and write this and the first two orders as

$$T = 1 - i \int J_\mu(1)\, a_\mu(1)\, d\tau_1 - \iint_{t_1 > t_2} V_{\mu\nu}(1, 2)\, a_\mu(1)\, a_\nu(2)\, d\tau_1\, d\tau_2 \qquad (2.2)$$

Evidently one can write an entire series of functions to ever increasing order.

<u>Unitarity 2nd Order.</u>

This gives the restriction (using $a^+ = a$)

$$- \iint_{t_1 > t_2} V^+_{\mu\nu}(1, 2)\, a_\nu(2)\, a_\mu(1)\, d\tau_1\, d\tau_2 + \iint_{\text{all } t} J_\mu(1)\, J_\nu(2)\, a_\mu(1)\, a_\nu(2)\, d\tau_1\, d\tau_2$$

$$- \iint_{t_1 > t_2} V_{\mu\nu}(1, 2)\, a_\mu(1)\, a_\nu(2)\, d\tau_1\, d\tau_2 = 0 \qquad (2.3)$$

obtained by writing $T^+T = 1$ and expanding to 2nd order. The 2nd integral is over all t_1, t_2. It can be split into a part $t_1 > t_2$ and a part $t_2 < t_1$ in the latter relable variables 1, 2 (and μ, ν) to get

$$+ \iint_{t_1 > t_2} \left[J_\mu(1)\, J_\nu(2) - V_{\mu\nu}(1, 2)\right] a_\mu(1)\, a_\nu(2)\, d\tau_1\, d\tau_2$$

$$+ \iint_{t_1 > t_2} \left[J_\nu(2)\, J_\mu(1) - V^+_{\mu\nu}(1, 2)\right] a_\nu(2)\, a_\mu(1)\, d\tau_1\, d\tau_2 = 0 \qquad (2.4)$$

Now, if we could assume that within the range of possible QED states arbitrary values for $a_\mu(1)\, a_\nu(2)$ can be generated (which is true) and also independently for $a_\nu(2)\, a_\mu(1)$ (which is false) we could conclude that

$$V_{\mu\nu}(1, 2) = J_\mu(1)\, J_\nu(2) \text{ for } t_1 > t_2 \qquad (2.5)$$

But outside the light cone for example $\left[a(2), a(1)\right] = 0$ so the two products are not independent, they are equal. To proceed therefore more carefully write

$$a_\nu(2)\ a_\mu(1) = a_\mu(1)\ a_\nu(2) + \left[a_\nu(2),\ a_\mu(1)\right]$$

to get

$$\iint_{t_1 > t_2} \left[J_\mu(1)\ J_\nu(2) + J_\nu(2)\ J_\mu(1) - V_{\mu\nu}(1,\ 2) - V^+_{\mu\nu}(1,\ 2)\right] a_\mu(1)\ a_\nu(2)\ d\tau_1\ d\tau_2$$

$$+ \iint_{t_1 > t_2} \left[J_\nu(2)\ J_\mu(1) - V^+_{\mu\nu}(1,\ 2)\right] \left[a_\nu(2),\ a_\mu(1)\right]\ d\tau_1\ d\tau_2 = 0 \quad (2.6)$$

Now the first factor must vanish, for we could take the case that a(2), a(1) commute first (for example one photon from an electron, another from a muon in lowest order) hence we surely always must have

$$J_\mu(1)\ J_\nu(2) + J_\nu(2)\ J_\mu(1) = V_{\mu\nu}(1,\ 2) + V^+_{\mu\nu}\ (1,\ 2) \qquad (2.7)$$

determining the real part of $V_{\mu\nu}(1,\ 2)$. In addition in general we must have (taking adjoint of last term)

$$\iint_{t_1 > t_2} \left[J_\mu(1)\ J_\nu(2) - V_{\mu\nu}(1,\ 2)\right] \left[a_\mu(1),\ a_\nu(2)\right] d\tau_1\ d\tau_2 = 0 \qquad (2.8)$$

The commutator is zero outside the light cone. Inside I believe we can make it arbitrary (although some little further study of special cases is necessary to verify this) hence we deduce

$$V_{\mu\nu}(1,\ 2) = J_\mu(1)\ J_\nu(2) \qquad \text{if}$$ 

(2.9)

1 is in forward light cone of 2.

We have almost proved Eq. (2.5) but not for any $t_1 > t_2$, only for t_1 inside the light cone of 2. The difference is very important because Eq. (2.5), to be relativistically invariant requires $\left[J_\mu(1),\ J_\nu(2)\right] = 0$ outside the light cone. Eq. (2.5) also is natural if hadrons come from any underlying field theory, for then our picture of coupling if $t_1 > t_2$ can

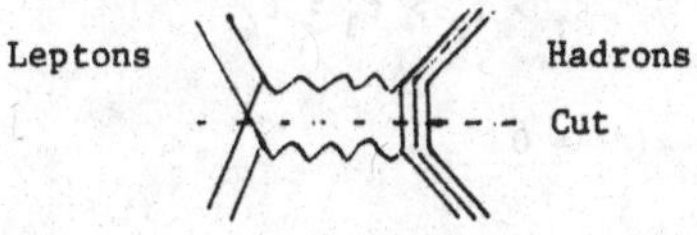

be cut at a t between t_2 and t_1; the first coupling is $J_\mu(1)$ and the second is $J_\nu(2)$, so we get the product. But one may be averse to assuming that strong interactions can be described by a complete set of states (and that the complete set can be taken as $|n, \text{in}\rangle$) at any arbitrary time. Nevertheless if we continue our study of the requirements for consistency with QED to 4th order we can do it.

Proof

This point is not important, but we do include the proof for completeness. If (2.5) were right then T would be $\left\{\exp - i\int J_\mu(1)\, a_\mu(1)\, d\tau_1\right\}_T$. In general it is some polynomial in $a_\mu(1)$, $\left\{a_\mu(2)\, a_\nu(1)\right\}_T$, $\left\{a_\mu(1)\, a_\nu(2)\, a_\sigma(3)\right\}_T$ etc. (From now on we omit the polarization indices - they always go in an obvious way with the position indices). The first order is made to agree with $J_\mu(1)$ so in general we can write (U(12) represents the deviation of V(12) from J(1) J(2)) for $t_1 > t_2$)

$$T = e^{-i\int J_\mu(1)\, a\,(1)\, d\tau_1}\Big[1 + i\int_{t_1}\int_{\geq t_2} U(1,\, 2)\, a(1)\, a(2)$$

$$+ \int_{t_1}\int_{\geq t_2}\int_{\geq t_3} U(1\;2\;3)\; a(1)\; a(2)\; a(3)$$

$$+ \int_{t_1}\int_{\geq t_2}\int_{\geq t_3}\int_{\geq t_4} U(1\;2\;3\;4)\; a(1)\; a(2)\; a(3)\; a(4) + \cdots)\Big] \qquad (2.10)$$

Now in forming T^+T to check unitarity the exp $-i\int Ja$ factors go out. To 2nd and 4th order in a the U (1 2 3) term does not enter. To second order we have

$$-\int U^+(1\;2)\; a(2)\cdot a(1) + \int U(1\;2)\; a(1)\; a(2) = 0$$

so we conclude U is hermitian $U^+ = U$ everywhere, and U(1 2) = 0 if 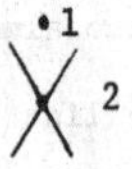

as before.

Now in 4th order we have

$$\int_{t_1 > t_2 > t_3 > t_4} U^+(1\;2\;3\;4)\; a(4)\; a(3)\; a(2)\; a(1) + \int_{t_1 > t_2 > t_3 > t_4} U(1\;2\;3\;4)\; a(1)\; a(2)\; a(3)\; a(4)$$

$$+ \int_{\substack{t_5 > t_6 \\ t_7 > t_8}} U(5\;6)\; U(7\;8)\; a(5)\; a(6)\; a(7)\; a(8) = 0 \qquad (2.11)$$

In the last integral we have $t_5 > t_6$ and $t_7 > t_8$ but no particular relation of t_5 and t_7 there are 6 relative orders. We replace in each the variables via t_1, t_2, t_3, t_4 in that order

1 .	5 .	5 .	5 .	. 7	. 7	. 7	call a(1) = A
2 .	6 .	. 7	. 7	5 .	5 .	. 8	a(2) = B
3 .	. 7	6 .	. 8	6 .	. 8	5 .	a(3) = C
4 .	. 8	. 8	6 .	. 8	6 .	6 .	a(4) = D.

The last term gives

$$U(12)\ U(34)\ a(1)\ a(2)\ a(3)\ a(4)\ :\ ABCD = ABCD$$

$$U(13)\ U(24)\ a(1)\ a(3)\ a(2)\ a(4)\ :\ ACBD = A[C, B]D + ABCD$$

$$U(14)\ U(23)\ a(1)\ a(4)\ a(2)\ a(3)\ :\ ADBC = A[D, B]C + AB[D, C] + ABCD$$

$$U(23)\ U(14)\ a(2)\ a(3)\ a(1)\ a(4)\ :\ BCAD = B[C, A]D + [B, A]CD + ABCD$$

$$U(24)\ U(13)\ a(2)\ a(4)\ a(1)\ a(3)\quad BDAC = B[D, A]C + [B, A]DC + AB[D, C] + ABCD$$

$$U(34)\ U(12)\ a(3)\ a(4)\ a(1)\ a(2)\quad CDAB = C[D, A]B + [C, A]DB + AC[D, B] + A[C, B]D + ABCD \qquad (2.12)$$

and the first term is

$$U^{+}(1\ 2\ 3\ 4)\ DCBA = [D, C]BA + C[D, B]A + CB[D, A] + [C, B]AD + B[C, A]D + [B, A]CD + ABCD\ . \qquad (2.13)$$

Now we get many obvious relations. For example the coefficient of ABCD must be zero (take case all 4 potentials commute). We believe that the vector potential a is an arbitrary function of space and time. We can therefore choose it to be different from zero only in four small regions of space time around the points 1, 2,3 and 4; call these regions σ_1, σ_2, σ_3, σ_4. For our special interest here take the case that the variables have the following light cone properties

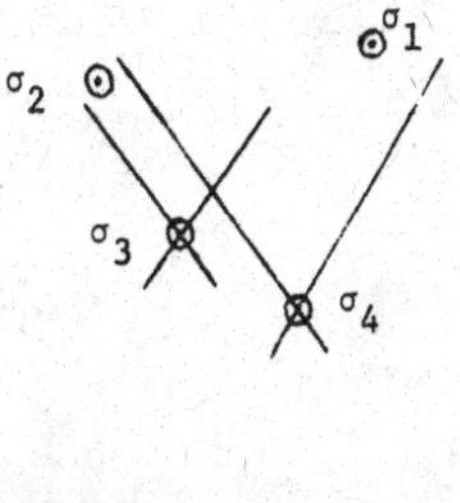

$$[A, C] = 0 = [A, B]$$
$$[B, D] = 0 = [D, C]$$
Only $[B, C]$ and $[A, D] \neq 0$

1 is outside the light cone of 2, 3 is outside the light cone of 4. Omitting the ABCD term which we noted was independently zero and collecting what is left in this case we get

$$\int_{\sigma_1+\sigma_2+\sigma_3+\sigma_4}\int\int\int \Big\{ U^{+}(1\ 2\ 3\ 4)\Big\{ [C,\ B]AD + CB[D,\ A]\Big\}$$
$$+ U(13)\ U(24)\ [C,\ B]AD$$
$$+ U(24)\ U(13)\ [D,\ A]BC$$
$$+ U(34)\ U(12)\Big\{[D,\ A]CB + [C,\ B]AD\Big\}\Big\}\, d\tau_1\, d\tau_2\, d\tau_3\, d\tau_4 \quad (2.14)$$

Only the third term needs to be turned around to $[D,\ A]CB + [D,\ A]\,[B,\ C]$ then all the terms are coefficients of $[C,\ B]AD$ or $CB[D,\ A]$ and must all vanish. But ultimately we are left with $U(24)\ U(13)\ [D,\ A]\,[B,\ C]$ or finally

$$\int_{\sigma_1+\sigma_2+\sigma_3+\sigma_4}\int\int\int U(24)\ U(13)\ \Big[a(2),\ a(3)\Big]\cdot\Big[a(4),\ a(1)\Big]\ d\tau_1\, d\tau_2\, d\tau_3\, d\tau_4 = 0\ . \quad (2.15)$$

Since the commutators are sufficiently general, I think we can conclude the integral will be zero only if the integrand is and $U(13) = 0$ even if 1 is outside the light cone of 3.

<u>End of Proof</u>

We can therefore conclude

$$V_{\mu\nu}\,(1,\ 2) = J_\mu(1)\ J_\nu(2) \qquad \text{for } t_1 > t_2\ . \quad (2.16)$$

In the proofs of (2.9) and (2.16) we have assumed that $a(\bar{x}_1,\ t_1)$ and $\big[a(\bar{x}_1,\ t_1),\ a(\bar{x}_2,\ t_2)\big]$ are sufficiently general functions of 1 and 2 (when 2 is in the light cone of 1). We believe this to be true. It is left to those interested in more rigorous proofs to verify this, for example by trying to construct a basis.

Lecture 3

We found in the previous lecture

$$V_{\mu\nu}(1,\ 2) = J_\mu(1)\ J_\nu(2) \qquad \text{for } t_1 > t_2 \quad (3.1)$$

furthermore

$$\left[J_\mu(1),\ J_\nu(2)\right] = 0 \qquad \text{for} \qquad$$ 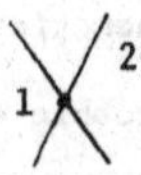

. (3.2)

This means the original symmetric $V_{\mu\nu}(1, 2)$ can be written

$$V_{\mu\nu}(1, 2) = \left\{J_\mu(1)\ J_\nu(2)\right\}_T + \text{seagulls}_{\mu\nu}\,\delta^4(1-2)\ . \tag{3.3}$$

The last term comes because there could be a $\delta(t_1 - t_2)$ term, or by relativity a $\delta^4(1-2)$ term or gradients thereof--leading to just a constant or polynomial under Fourier transform.

These seagull terms would mean that the abstract form for T would be like

$$\left\{e^{-i\left[\int J_\mu(1)a_\mu(1)d\tau + \int S_{\mu\nu}(1)a_\mu(1)a_\nu(1)d\tau_1 + \dots\right]}\right\}_T$$

thus adding a local term at one space time point, but second order in a(1) as appears in QED for the interaction with a scalar particle, for example. Of course, instead it could contain gradients, as $F_{\mu\nu}(1)\ F_{\mu\nu}(1)$ for example. There can be higher terms for higher orders. In short what we have found is that T must be expressible in the form $T = \left\{\exp\left[-i\int L(1)d\tau_1\right]\right\}_T$ where L(1) is an operator depending on a(1) only, for example of form $J_\mu(1)\ a_\mu(1) + {}$ $+S_{\mu\nu}(1)\ a_\mu(1)\ a_\nu(1) + \dots$ where J , S are operators in the hadron variables. Such a form is of course, also, the immediate result of supposing a local field theory for hadrons.

Research Problem

What experiments could best establish existence or non-existence of seagulls?

In QED for spin 1/2 there is no seagull, for spin 0 there is (but with Kemmer Duffin matrices there is not – resolve this!) Since all quantities are defined by experiment the reality of such seagulls for hadrons is an experimentally determinable fact.

Thus we see a knowledge of matrix elements of J_μ alone should determine all scattering amplitudes V.

The restriction (3.2) on the matrix elements of J_μ are very important. They lead to many relations--dispersion relations for example. We return to this subject later when we discuss deep inelastic ep scattering. There are some special technical difficulties coming from the highly divergent nature of some of these expressions so mathematical rigor requires a little more attention. In practice they give trouble only in the vacuum expectation (of, for example $V_{\mu\nu}$) and in no other problem and so they can be avoided best by disregarding them. (They have been analyzed by Schwinger, and are called trivial Schwinger terms).

Conservation of Current

We suppose now that current is conserved in the more conventional sense that we will take it to be true that quantum electrodynamics cannot determine $a_\mu(1)$ completely, but a gradient $a_\mu(1) + \nabla_\mu \chi(1)$ ($\chi(1)$ is an arbitrary function, not an operator) can be added to it (i.e. if a different gauge were used in the lepton theory) without altering the physics. Thus $T[a] = T[a + \nabla\chi]$ so

$$1 - i\int J_\mu(1)\, a_\mu(1)\, d\tau_1 - \frac{1}{2}\int V_{\mu\nu}(12) \left\{a_\mu(1)\, a_\nu(2)\right\}_T d\tau_1\, d\tau_2 + \ldots$$

$$= 1 - i\int J_\mu(1) \left(a_\mu(1) + \nabla_\mu\left(\chi(1)\right)\right) d\tau_1 - \frac{1}{2}\int V_{\mu\nu}(12) \left\{\left(a_\mu(1) + \nabla_\mu\chi(1)\right) \left(a_\nu(2) + \nabla_\nu\chi(2)\right)\right\} d\tau_1 d\tau_2 + \ldots \tag{3.4}$$

or to first order in χ

$$\nabla_\mu J_\mu(1) = 0 \tag{3.5}$$

$$\nabla_\mu^{(1)} V_{\mu\nu}(1, 2) = 0 \qquad \text{(3.6) etc.}$$

Equation (3.5) we have already discussed; (3.6) gives something new. Using (3.3)

$$\nabla_\mu V_{\mu\nu}(1, 2) = \nabla_\mu \left\{J_\mu(1)\, J_\nu(2)\right\}_T + \nabla_\mu\, \text{seagulls}_{\mu\nu}\, \delta^4(1, 2)$$

but

$$\nabla_\mu \left\{J_\mu(1)\, J_\nu(2)\right\}_T = \left\{\nabla_\mu J_\mu(1)\, J_\nu(2)\right\}_T = 0$$

by equation (3.5; except that ∇_μ does not commute with the time ordering operation in

$$\left\{J_\mu(1)\, J_\nu(2)\right\}_T = \theta(t_1 - t_2)\, J_\mu(1)\, J_\nu(2) + \theta(t_2 - t_1)\, J_\nu(2)\, J_\mu(1).$$

We have also to differentiate the θ with respect to t. Thus when $\mu = 0$, we get from this an extra term

$$\delta(t_1-t_2)\left[J_o(1),\, J_\nu(2)\right] ;$$

the equal time commutator of $J_o(1)$ and $J_\nu(2)$. Thus in general

$$\nabla_\mu V_{\mu\nu}(1,\, 2) = \left\{\nabla_\mu J_\mu(1),\, J_\mu(2)\right\} + \delta(t_1-t_2)\left[J_o(1),\, J_\nu(2)\right] + \\ + \nabla_\mu \text{ seagulls}_{\mu\nu}\ \delta^4(1-2) .$$

In our application this leads to

$$\delta(t_1-t_2)\left[J_o(1),\, J_\nu(2)\right] = -\nabla_\mu \text{ seagulls}_{\mu\nu}\ \delta^3(\bar{x}_1-\bar{x}_2)\delta(t_1-t_2)$$

or $$\left[J_o(1),\, J_\nu(2)\right]_{t_1=t_2} = -\nabla_\mu \text{ seagulls}_{\mu\nu}\ \delta^3(\bar{x}_1-\bar{x}_2) \qquad (3.7)$$

Actually I have simplified the discussion of the seagulls, for gradients also appear on the $\delta^4(1,\, 2)$, but the point we want to make is that the equal time commutator of charge density and current is determined by seagulls. In particular, if as many people (e.g. Gell-Mann) have suggested seagulls vanish (by analogy to QED where the coupling is purely $\int\bar{\psi}A_\mu\gamma_\mu\psi d\tau$ to a spin 1/2 field there are no seagull diagrams) we would have

$$\left[J_o(1),\, J_\nu(2)\right]_{t_1=t_2} = 0 \qquad (3.8)$$

we have no direct test of this yet (although we do have tests of $\nabla_\mu V_{\mu\nu}(1,\, 2) = 0$).

__Remark__

Schwinger has pointed out that (3.8) is impossible. Because if $J_o = \rho$ and $J_\nu = 1, 2, 3$ is written on the vector $\bar{J}$ it says

$$\left[\rho(\bar{x}_1),\, \bar{J}(\bar{x}_2)\right] = 0$$

Therefore taking divergence $\left[\rho(\bar{x}_1),\, \nabla_2\cdot\bar{J}(\bar{x}_2)\right] = 0$ or $\left[\rho(\bar{x}_1),\, \frac{\partial\rho}{\partial t}(\bar{x}_2)\right] = 0$ because $\nabla\cdot\bar{J} = \frac{\partial\rho}{\partial t}$. But $\frac{\partial\rho}{\partial t}$ is the operator $H\rho-\rho H$ where H is the Hamiltonian of the system (assuming there is one—generally the energy of state operator) or

$$\rho H\rho - \rho\rho H = 0$$

Now take the vacuum expectation value. Let the energy of the intermediate state n be En, we have

$$\sum_n(\rho_{on}E_n\rho_{no} - \rho_{on}\rho_{no}E_o) = 0,$$

but $\rho_{on} = \rho^*_{no}$ and if the vacuum is the lowest state we must have $E_n - E_o > 0$ so

$$\sum_n(E_n - E_o)^2|\rho_{no}|^2 = 0$$

which is impossible.

But this argument applies as well to QED itself where we know we have no seagulls in the original field operator for the lagrangian. It comes because we do not really compare this formal field theory directly to experiment but remove some divergent vacuum diagrams at the beginning. This problem appears entirely associated with the vacuum problem and could be removed. We can indeed have the analogy of QED; 'no seagulls' and have equation (3.8) satisfied for every problem except the vacuum. The precise statement is that (3.8) holds if from the commutator you subtract its vacuum expectation value times a unit matrix.

Lecture 4

Isotopic Spin, Strangeness, Generalized Currents

The hadronic states n, m of a matrix element of J_μ such as $\langle m|J_\mu|n\rangle$ can be classified into definite nondynamical quantum numbers of isospin and strangeness (both of which, we assume, are perfectly conserved by the strongly interacting system above). The J matrix may have elements between different values of the quantum numbers, but it must of course conserve charge. From the very low rate of $K^o \to \pi^o + \gamma$ or $\Lambda \to \gamma + n$ (although $\Sigma^o \to n + \gamma$ is fast enough) we conclude that weak interactions are involved here. Thus we think J has zero matrix elements between states of different strangeness. Among sets of different isospin we can describe the result by saying J has parts of I spin = 0 (isoscalar), I = 1 (isovector), I = 2 (isotensor) etc. and use appropriate Clebsch Gordan coefficients to relate amplitudes among different multiplets. Since J does not change charge it must only involve the 3rd component, if it is isovector for example. The fact that proton and neutron have charges + 1, 0 already shows that J is not pure isoscalar independent of isospin, nor pure isovector (where the charges would have to be opposite) but contains a linear combination of these two. No experiment seems to require I = 2, but I do not know how precisely or extensively this has been tested. Recently some evidence was claimed for the need of an I = 2 component in comparing $\gamma p \to \pi^+ N$ and $\gamma N \to \pi^- p$ at energies near the Δ resonance--but it appears that corrections for deuteron structure (for the γ neutron rate is inferred from γD data) were incorrectly analyzed.

Most theorists today assume $\Delta I = 0$ or $\Delta I = 1$ only for J_μ. (This is evidently a fundamental question because it tells something of how J is "ultimately" coupled; for further strong interactions conserving I-spin cannot alter this rule--we see "in" through the strong dynamic coupling in this respect at least because the strong coupling conserves this I spin character.)

Having available matrix elements $\langle m|J_\mu|n\rangle$ for a variety of states n (and m) all of the same I-spin multiplet permits one by appropriate linear combination always to isolate the pieces due to the isoscalar and isovector part separately. Thus we can define matrix elements and therefore operators for $J_\mu^S(q)$ and $J_\mu^{V3}(q)$. But for the vector we could also calculate (via Clebsch Gordan coefficients) matrix elements between specific states of other components of the vector current $J_\mu^{V+}(q)$ or $J_\mu^{V-}(q)$ (with isospin +1 or -1). In this way new kinds of currents are definable.

This would just be an exercise in Clebsch Gordan coefficients, but we think some of these currents are physically important also. We think the current $J_\mu^{V+}(q)$ is the nonstrangeness changing nonparity violating part of weak interaction (an assumption known as CVC). This leads to a suggestion by which these extensions of current are useful in a powerful theoretical way (Gell-Mann).

That QED is coupled to J_μ, or weak interaction to J_μ^{V+} is (in our point of view in these lectures) a kind of accident irrelevant to strong interactions. They just lead us to a tool to study hadrons, but hadron interactions can be analyzed alone. Nevertheless we deduced a number of relations from assuming either that these currents come from some operator in a field theory underlying hadrons, or that hadrons are such that weak perturbation fields could be coupled (we will use the latter hypothesis). We can expect, for any two points and components of current $J_\mu^a(1)$ and $J_\nu^b(2)$ that they commute if 1, 2 are space like separated

$$\left[J_\mu^a(1),\, J_\nu^b(2)\right] = 0. \qquad \text{if } 2 \times \bullet 1 \tag{4.1}$$

This is a new assumption. We are trying to induce new laws and restrictions on J and the hadron system. We know $\left[J_\mu(1),\, J_\nu(2)\right] = 0$ for $2 \times \bullet 1$ where J, our electromagnetic current, is an isovector and isoscalar. Using isospin only how far is it possible to go to prove say that the isovector part, or the $\left[J_\nu^+(1),\, J_\mu^+(2)\right]$ etc. commute? In the realm of isospin what we assume here is that the space-like commutation law is true not only for the total current $J^S + J^{V_3}$ but also for the isoscalar part alone with itself, the isovector part alone with itself, the isoscalar part with the isovector part and generalizations for the isovector part with different I spin component directions.

We can also have the scattering of an imaginary a-type vector "photon" to a b-type photon governed by

$$V_{\mu\nu}^{ab}(1, 2).$$

That is, extend the concept of vector potential $a_\mu(1)$ to contain another index a, the type, carrying for example isospin or strangeness. Then the coupling to external potentials could be

$$T = 1 - i \int J_\mu^a(1)\, a_\mu^a(1)\, d\tau_1 - \frac{1}{2} \iint V_{\mu\nu}^{ab}(1\ 2) \left\{a_\mu^a(1)\, a_\nu^b(2)\right\}_T d\tau_1\, d\tau_2 + \dots \tag{4.2}$$

$$\text{Thus} \quad V_{\mu\nu}^{ab}(1, 2) = \left\{J_\mu^a(1)\, J_\nu^b(2)\right\}_T + \text{seagulls}_{\mu\nu}^{ab}\, \delta^4(1-2) \tag{4.3}$$

<u>Conservation of Generalized Currents</u>

We want the conservation laws analogous to $\nabla_\mu V_{\mu\nu}(1\ 2) = 0$. Take first the case if I spin which we know is exactly conserved. Consider the scattering of an I = 1 particle via J_μ^{V+}. The charge of the final state is one higher than the initial, so

$$\left[Q,\, J_\mu^{V+}(1)\right] = J_\mu^{V+}(1) \tag{4.4}$$

But $Q = \int J_o^{V3}(1)\, d^3\bar{x}$ so

$$\left[\int J_o^{V3}(t_1, \bar{x})\, d^3\bar{x},\ J_\mu^{V+}(t_1, \bar{x}')\ \right] = J_\mu^+(t_1, \bar{x}')$$

$$= \int d^3\bar{x} \left[J_o^{V3}\,(t_1, \bar{x}),\ J_\mu^{V+}(t_1, \bar{x}')\right] \tag{4.5}$$

but the last, equal time commutator is by hypothesis (4.1) zero outside the light cone—it must be a multiple of $\delta^3(\bar{x}-\bar{x}')$, say $s(\bar{x})\delta^3(\bar{x}-\bar{x}')$. Equation (4.1) shows s must be $J_\mu^{V+}(\bar{x})$. Hence the equal time commutation relation results,

$$\left[J_o^{V3}(t_1, \bar{x}_1),\ J_\mu^{V+}(t_1, \bar{x}_2)\right] = \delta^3(\bar{x}_1-\bar{x}_2)\ J^{V+}(t_1, \bar{x}_1) \tag{4.6}$$

(this assumes there are no special terms in $\delta(\bar{x}_1, \bar{x}_2)$ which would integrate out, a question we shall see related to seagulls again).

Equation (4.6) and its generalization to the much wider group $SU_3 \times SU_3$ are Gell-Mann's equal time commutator relations. They represent the first guessed dynamical property of hadrons that is not simply a consequence of relativistic quantum mechanics general principles.

We can also describe this from the point of view of a property of the scattering function $V_{\mu\nu}^{ab}$. Take the case one potential is + isospin, the other I_3. Since current is conserved you might at first think $\nabla_{\mu 1}\ V_{\mu\nu}^{3+}(1, 2) = 0$, but the error is that the potential ν, 2 carries in a charge +. So only if the electromagnetic potential (coupling 3) also couples to the + meson is charge conserved. E.g. diagrams are like

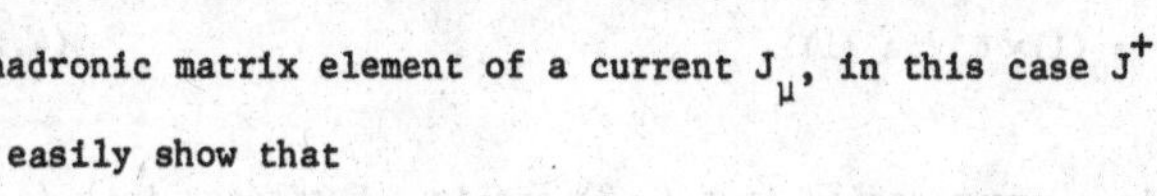

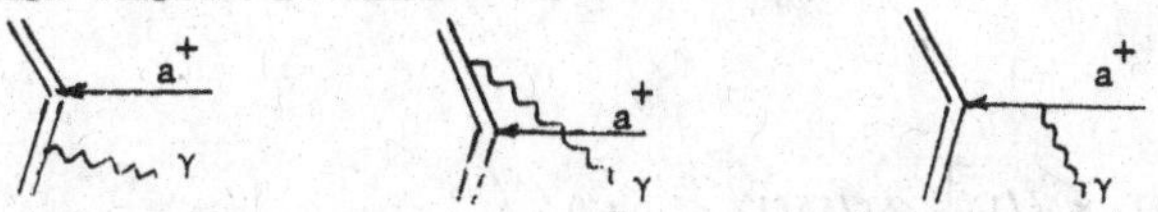

Their sum conserves charge, and if the photon had polarization $e_\mu = \alpha q_\mu$ proportional to its own momentum the sum would give zero. $V_{\mu\nu}^{3+}(1, 2)$ is just the sum of the first two, the last one is easily computable and is clearly a first order hadronic matrix element of a current J_μ, in this case J^+ itself. Thus one can easily show that

$$\nabla_\mu J_\mu^+(1) = 0 \qquad \text{(just by I-spin from } \nabla_\mu J_\mu^3(1) = 0\text{)}$$

$$\nabla_\mu V_{\mu\nu}^{3+}(1, 2) = \delta^4(1-2)\ J_\nu^+(2)\ . \tag{4.7}$$

As we shall show in a moment (4.7) is equivalent to (4.6) if no seagulls exist; if they do exist (4.7) is true but (4.6) has to be modified. (4.7) is the more fundamental relation.

To show the relation of (4.7) to (4.6) we substitute (4.3) into (4.7).

$$\nabla_\mu \left\{ J^3_\mu(1),\ J^+_\nu(2) \right\}_T =$$

$$\left\{ \nabla_\mu J^3_\mu(1),\ J^+_\nu(2) \right\}_T + \delta(t_1-t_2) \left[J_o(t_1, \bar{x}_1),\ J^+_\nu(2) \right]_{t_1=t_2} \tag{4.8}$$

so writing $\delta^4(1\text{-}2)$ on the right side of (4.7) or $\delta(t_1-t_2)\ \delta^3(\bar{x}_1-\bar{x}_2)$ we obtain the result (4.6) since $\nabla_\mu J^3_\mu = 0$. In general seagull terms must be added, but we do not know if they exist.

The generalization to a general Lie group with generators G^a with commutation relation $[G^a, G^b] = f^c_{ab}\ G^c$ ($= G^{a\times b}$ by definition) $(a\times b)_c = f^c_{ab}$ is

$$\nabla_\mu J^a_\mu(1) = 0 \tag{4.9}$$

$$\nabla_{\mu 1} V^{ab}_{\mu\nu}(1, 2) = \delta^4(1\text{-}2)\ J^{a\times b}_\nu(1) \tag{4.10}$$

These may be obtained from noting that the generalization of a gauge transformation $a'_\mu \to a_\mu + \nabla_\mu \chi$ is, in the group $a^a_\mu \to a^a_\mu + \nabla_\mu \chi^a + (\chi \times a_\mu)^a$. Supposing $T[a]$ is unchanged by such a transformation we find

$$T\left[a + \nabla\chi + \chi \times a\right] = T[a]$$

as a functional relation—or calling $\delta T/\delta a(1)$ the functional derivative one easily deduces

$$\int \left[\nabla\chi(1) + \chi(1) \times a(1)\right]\ \delta T/\delta a(1)\ d\tau_1 = 0$$

for all $\chi(1)$, so integrating by parts we have

$$\nabla_\mu \delta T/\delta a^a_\mu(1) = a_\mu(1) \times \delta T/\delta a_\mu(1)\ . \tag{4.11}$$

When T is written in a power series in a (4.2) and is substituted into (4.11), zero and first order terms give (4.9) and (4.10).

Since isospin is exactly conserved (4.9), (4.10) must be exactly satisfied when restricted to the three spin components of vector currents. What of SU_3 which is only "almost" satisfied? Gell-Mann has proposed that SU_3, although not exactly satisfied for the entire hadronic system may be more and more accurately satisfied as shorter and shorter space-time intervals are involved. That is how it would behave if an underlying field theory had propagator gradient terms satisfying SU_3, but mass-like terms violating SU_3. (E.g. $\bar{q}\not\nabla q + \bar{q}mq$ where m is a nonSU_3 invariant matrix, q are quark operators). If J^a is a strangeness changing current, having for example a matrix element between Λ, N then $\nabla_\mu J^a_\mu(1) \neq 0$ because Λ, N have different masses. (If Λ, N at rest $\langle\Lambda|J_{xyz}|N\rangle = 0$, $\langle\Lambda|J^a_o|N\rangle = \alpha$ say, therefore $\langle\Lambda|\frac{\partial J^a_o}{\partial t}|N\rangle - \langle\Lambda|\nabla\cdot J|N\rangle = (\omega_\Lambda - \omega_N)\alpha$ cannot be zero.) That is $\nabla_\mu J^a_\mu(1)$ is equivalent to another operator, say $n^a(1)$. Then

$$\nabla_\mu V^{ab}_{\mu\nu}(1, 2) = \delta(t_1-t_2)\left[J^a_o, J^b_\nu\right]_{t_1=t_2} + \left\{n^a(1)\ J^b_\nu(2)\right\}_T \tag{4.12}$$

Now the latter presumably does not contain a singularity as strong as $\delta^4(1-2)$ but if SU_3 is valid at small enough distances—let us say the $\delta^4(1-2)$ singularity of $\nabla_\mu V^{ab}_{\mu\nu}$ is correctly given by (4.10). Thus we say

$$\nabla_\mu V^{ab}_{\mu\nu}(1, 2) = \delta^4(1-2)\ J^{axb}_\nu(2) + \text{"smooth"} \tag{4.13}$$

where "smooth" is less singular than $\delta^4(1-2)$. Then we can still deduce the equal time commutator relations under the above assumptions of smoothness of the SU_3 violating term. Equating the singular terms in (4.12) and (4.12) we find

$$\left[J^a_o, J^b_\nu\right]_{t_1=t_2} = \delta^3(\bar{x}_1-\bar{x}_2)\ J^{axb}_\nu(2) \tag{4.14}$$

(seagull terms have been ignored).

These relations are of very great interest because they are nonlinear requiring absolute scales. Thus (if valid) they can serve as supplying absolute scale definitions to the currents so that the rule that weak interaction of hadrons is V + A (rather than V - .7A) is definable and therefore testable. This particular test has been made by Adler and Weisberger using PCAC to take the pion coupling as a measure of the divergence of the axial current. We discuss how somewhat more direct tests can be made by neutrino scattering later on in the course (in Part II).

Singularities on the Light Cone

The commutator $\left[J_\mu(1), J_\nu(2)\right]$ is zero if 1 is outside the light cone of

2, nonzero inside. What kind of singularity does it have just passing through, or near the light cone. For free fields of any mass in field theory the commutator has a $\delta(s_{12}^2)$ type singularity across the cone. There is experimental evidence from inelastic scattering experiments of electrons on protons that the singularity of $\left[J_\mu(1),\ J_\nu(2)\right]$ is of the same type (gradients of) $\delta(s_{12}^2)$ where s_{12} is the interval from 1 to 2. We shall discuss this matter in considerable detail where we discuss these experiments. At this time we shall also develop further other formal properties of the commutator or time ordered products of J_μ operators. These matters will therefore be deferred until later, for I find them easier to discuss when certain experiments are in mind.

Vacuum Expectation of $V_{\mu\nu}(1, 2)$

In order, however, not to leave this theoretical discussion entirely in the air, I shall illustrate one application of it--the simplest, the vacuum expectation of $V_{\mu\nu}(1, 2)$. This is a function only of the difference 1 - 2. Its fourier transform, into variable q--which we will call $V_{\mu\nu}(q)$ is needed, for example, to calculate vacuum polarization corrections (to order e^2) due to hadrons. It represents the diagram

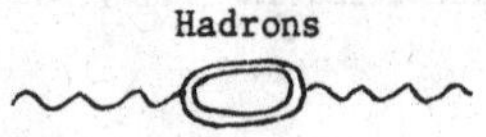

If we write in momentum space $\langle 0|V_{\mu\nu}(-q, q)|0\rangle = -q_\mu q_\nu\ v(q^2) + \delta_{\mu\nu} b(q^2)$ (by relativistic invariance, gauge invariance implies $b = q^2 v$ or we have

$$\langle 0|V_{\mu\nu}(-q, q)|0\rangle = (\delta_{\mu\nu} q^2 - q_\mu q_\nu)\ v(q^2). \tag{4.15}$$

If we write $(\delta_{\mu\nu} - \frac{q_\mu q_\nu}{q^2})b$ we see b must go to zero as $q \to 0$ in order to avoid a new pole at $q^2 = 0$. Acting on conserved currents the last term vanishes.

The series of bubbles propagating between two currents is

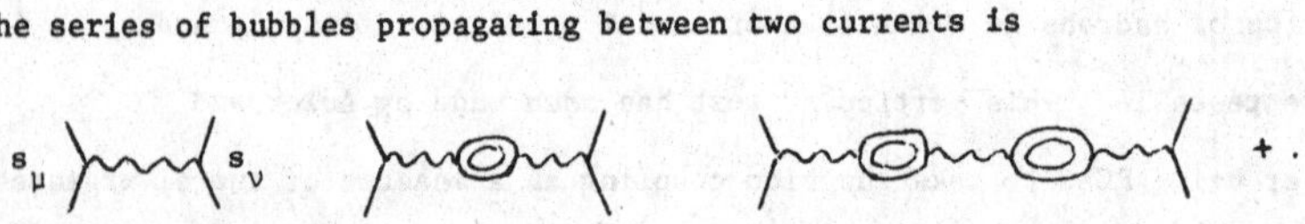

$+ \ldots$

$$= s_\mu \frac{4\pi e^2 i}{q^2} s_\nu + s_\mu \frac{4\pi e^2 i}{q^2} \cdot q^2 v(q^2) \frac{4\pi e^2 i}{q^2} s_\nu + s_\mu \frac{4\pi e^2 i}{q^2} q^2 v \frac{4\pi e^2 i}{q^2} q^2 v \frac{4\pi e^2 i}{q^2} s_\nu + \ldots$$

$$= 4\pi e^2\ s_\mu \frac{1}{q^2(1-4\pi e^2 i v(q^2))} s_\nu\ . \tag{4.16}$$

We note there is no mass renormalization of the photon, the pole is still at $q^2 = 0$, but the residue is altered to $\frac{4\pi e^2}{1-4\pi e^2 ia}$ where $a = v(0)$ (which may turn out infinite) is lost in charge renormalization. If we start expanding near $q^2 = 0$ to get $v(q^2) = a + q^2 b$, we can write the renormalized propagator to first order in e^2 as

$$\frac{1}{(q^2 - 4\pi e^2 ibq^4)}$$

so $4\pi e^2 ib$ measures the vacuum polarization correction due to hadrons in such predominantly low energy QED problems as the Lamb Shift etc.

The imaginary part of $v(q^2)$ for $q^2 > 0$ is the "virtual photon lifetime" and gives the rate of production of hadrons in (say) an electron-position collision. Because the imaginary part of the amplitude represents a loss in probability that a photon remains a photon

 $= 1 + 4\pi e^2 iv(q^2)$ i.e. Prob. $= 1 + 4\pi e^2 i(v - v^*)$.

It is therefore directly accessible to experiment. The real part is related to the imaginary part by a dispersion relation. Therefore hadronic vacuum polarization effects (to order e^2) could be completely determined after suitable experiments are done. We will discuss this matter in detail in the next lecture.

Lecture 5

$e^+ + e^- \rightarrow$ Any Hadrons

Consider the process $e^+ + e^- \rightarrow$ hadrons in state m out. It is governed by an amplitude

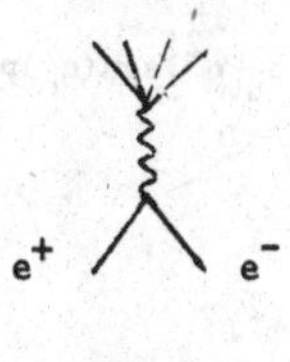

$$(\bar{u}_2 \gamma_\mu u_1) \frac{4\pi e^2 i}{q^2} \langle m_{out} | J_\mu(q) | 0 \rangle \qquad (5.1)$$

Hence the probability is proportional to

$$|<m_{out}|J_\mu(q)|0>|^2 = <0|J_\mu(-q)|m_{out}><m_{out}|J_\mu(q)|0>$$

Thus if we could measure the total cross section for $e^+ + e^- \to$ hadrons in any state as a function of the energy E of the electron or position in the case ($q_0 = 2E$, $\bar{Q} = 0$, $q^2 = 4E^2$) we can directly measure $\Sigma m_{out} <0|J_\mu(-q)|m_{out}><m_{out}|J_\mu(q)|0>$

$$\sum_{m\ out} <0|J_\mu(-q)|m_{out}><m_{out}|J_\nu(q)|0> =$$

$$<0|J_\mu(-q)J_\nu(q)|0> = p_{\mu\nu}(q) \tag{5.2}$$

Relativity and gauge invariance permits us to write this in the form $(q_\mu q_\nu - q^2\delta_{\mu\nu})\,\theta(q_0)p(q^2)$ because we know that if $q_0 < 0$ (the vacuum state being lowest) no hadron state m_{out} could be excited. $p_{\mu\nu}(q) = 0$ for $q_0 < 0$. (In fact if Q, the space like momentum of q is nonzero, the lowest possible m_{out} state is a pair of pions with momentum Q hence $p_{\mu\nu}(q) = 0$ unless $q_0 > 2\sqrt{m_\pi^2 + Q^2/4}$ or $q^2 > (2m_\pi)^2$ and $q_0 > 0$.) Thus $p(q^2)$ exists only for $q^2 > (2m_\pi)^2$. For example, $p(q^2) = 0$ for $q^2 < 0$, i.e., space like q^2. Further $p(q^2)$ must be positive, for example if $q = (q_o, 0, 0, 0)$ and the polarization of the virtual photon were x, the sum of absolute squares must be positive on the right hand side of (5.2). Therefore $p(q^2)$ is positive for $q^2 > (2m_\pi)^2$, zero if $q^2 < (2m_\pi)^2$.

Therefore

$$<0|J_\nu(1)J_\mu(2)|0> \overset{F.T.}{=} <0|J_\nu(-q)J_\mu(q)|0> = (q_\mu q_\nu - \delta_{\mu\nu}q^2)\theta(q_0)p(q^2) \tag{5.3}$$

$$<0|J_\mu(2)J_\nu(1)|0> \overset{F.T.}{=} <0|J_\mu(q)J_\nu(-q)|0> = (q_\mu q_\nu - \delta_{\mu\nu}q^2)\theta(-q_0)p(q^2) \tag{5.4}$$

(F. T. stands for Fourier transform).

Now we can work out the commutator and time ordered product. First the commutator

$$<0|\Big[J_\nu(1),\ J_\mu(2)\Big]|0> = (5.3)-(5.4) = (q_\mu q_\nu - \delta_{\mu\nu}q^2)\mathrm{sgn}(q_0)p(q^2). \tag{5.5}$$

Note this can be written as

$$(q_\mu q_\nu - \delta_{\mu\nu}q^2)\mathrm{sgn}(q_0) \cdot \int_{(2m_\pi)^2}^{\infty} \delta(q^2-m^2)p(m^2)dm^2. \tag{5.6}$$

Therefore the vacuum expectation of the commutator in space is

$$(\nabla_{\mu 1}\nabla_{\nu 1} - \delta_{\mu\nu}\nabla_1^2) \int_{(2m_\pi)^2}^{\infty} dm^2\, p(m^2) \cdot C^m(1,2), \tag{5.7}$$

which is zero outside the light cone, because $C^m(1,2)$ is. It is interesting that the proof that the commutator vanishes outside the light cone, assuming relativity is so simple for the vacuum expectation. (At the end of these notes the commutator, propagator and their Fourier transforms are listed). To get the time ordered product we need

$$<0|\{J_\nu(1)J_\mu(2)\}_T|0> = <0|1(t_1-t_2)J_\nu(1)J_\mu(2) + 1(t_2-t_1)J_\mu(2)J_\nu(1)|0>$$

To get the F. T. we need a convolution with the F. T. of $\theta(t_1-t_2)$

$$\int \theta(t)e^{+i(q_o t-\bar{Q}_o\bar{x})} d^3\bar{x}dt = (2\pi)^3\delta^3(\bar{Q}) \frac{i}{q_o+i\varepsilon}$$

$$\int \theta(-t)e^{+i(q_o t-\bar{Q}_o\bar{x})} d^3\bar{x}dt = (2\pi)^3\delta^3(\bar{Q}) \frac{-i}{q_o+i\varepsilon}$$

Therefore

$$\text{F. T. } <0|\{J_\nu(1)J_\mu(2)\}_T|0> = \frac{i}{2\pi}\int\left(\frac{1}{q_o-q_o'+i\varepsilon}\,\theta(q_o')\right.$$

$$\left. - \frac{1}{q_o-q_o'-i\varepsilon}\,\theta(-q_o')\right)(q_\mu'q_\nu' - \delta_{\mu\nu}q'^2)p(q'^2)dq_o' = V_{\mu\nu}(q) \tag{5.8}$$

where q' means $(q_o', \bar{Q})$. First do the case $\mu,\nu = t,t$ for then $q_\mu'q_\nu' - \delta_{\mu\nu}q'^2 = Q^2 = q_t q_t - \delta_{tt}q^2$ and comes out of the integral. Then change the sign of q_o' in the second term integral to get

$$<0|\{J_t(1)J_t(2)\}|0> \overset{\text{F.T.}}{=} \frac{i}{2\pi}(q_t q_t - \delta_{tt}q^2)\int p(q'^2)\theta(q_o')\left(\frac{1}{q_o-q_o'+i\varepsilon}\right.$$

$$\left. - \frac{1}{q_o+q_o'-i\varepsilon}\right) dq_o' = V_{tt}(q).$$

The integral is just

$$\frac{i}{2\pi}\int p(q'^2)\theta(q_o')2q_o'dq_o' \frac{1}{q_o-q_o'^2+i\varepsilon} ,$$

or changing variable from q_o' to $q_o'^2 - Q^2 = m^2$ get

$$V_{tt}(q) = (q_t q_t - \delta_{tt}q^2)\int \frac{p(m^2)dm^2}{q^2-m^2+i\varepsilon} - iS_{tt} \tag{5.9}$$

where S_{tt} is the F. T. of a possible seagull expectation in the vacuum. It must be a constant, or a finite polynomial in q, and must be real. We can guess (generalizing $\mu\nu$ from tt--however see note below) that writing $V_{\mu\nu}(q) = (q_\mu q_\nu - \delta_{\mu\nu} q^2)\ v(q^2)$ we have,

$$v(q^2) = \frac{i}{2\pi} \int_{(2m_\pi)^2}^{\infty} \frac{p(m^2)dm^2}{q^2-m^2+i\varepsilon} - iS \tag{5.10}$$

where S, at worst, is a finite polynomial in q; at best it is zero. Note the imaginary part of $iv(q^2) = \frac{1}{2}\,p(q^2)$ as previously remarked physically we can also write

$$iv(q^2) = -\frac{1}{\pi} \int \frac{\mathrm{Im}\left(iv(q^2)\right)_{\text{at } m^2}}{q^2-m^2+i\varepsilon}\, dm^2 + S \quad . \tag{5.11}$$

This is a dispersion relation for $v(q^2)$, expressing v for all q^2 in terms of its imaginary part (which in this case is (a) nonzero only for $q^2 > (2m_\pi)^2$, (b) measurable by accessible experiments.)

If $\sigma(s)$ is the cross section for annihilation of electron-positron into any hadrons where s is the total C.m. energy squared, we have $\sigma(s) = (4\pi e^2)^2 p(s)/2$ so the vacuum polarization is given directly in terms of experiment by

$$4\pi e^2 i\left[v(q^2) - v(0)\right] = \frac{q^2}{\pi(4\pi e^2)} \int_{4m_\pi^2}^{\infty} \frac{\sigma(s)ds}{s-q^2} \quad . \tag{5.13}$$

The Lamb shift correction (or correction to the magnetic moment of the electron, etc.) due to hadron vacuum polarization depends on $4\pi e^2 iv'(0)$ which is

$$\frac{1}{4\pi^2 e^2} \int \frac{\sigma(s)ds}{s} \quad .$$

It is expected today for reasons to be discussed later that $\sigma(s)$ may, for large s vary as constant/s so the integral would converge and be determined experimentally.

In general the possibility of an S must be resolved to get the greatest use of such a relation. Otherwise if $v(q^2)$ is known somewhere at some special value of q^2 (say 0 or ∞) we can convert to a subtracted dispersion relation. I.e. suppose it is known S is a constant, but not a polynomial in q^2, and $v(q_1^2)$ is known. Subtract from (5.10) for q^2 its value when $q^2 = q_1^2$ to get

$$v(q^2) - v(q_1^2) = \frac{i}{2\pi} \int \frac{p(m^2)dm^2(q_1^2-q^2)}{(q^2-m^2+i\varepsilon)(q_1^2-m^2+i\varepsilon)} \tag{5.12}$$

S does not appear. This same trick helps if the integral on m^2 looks divergent.

The integral on m^2 may now converge better for large m^2.

In our application any constant S is uninteresting, as we have seen the value of v(0) is of no interest today because it is lost in charge renormalization. Hence supposing S does not have a q^2 term (assumption of no bad seagulls) we would write a dispersion relation for the quantity $v(q^2) - v(0)$, the only quantity of physical significance

$$v(q^2) - v(0) = \frac{iq^2}{2\pi} \cdot \int_{(2m_\pi)^2}^{\infty} \frac{\rho(m^2)dm^2}{(q^2-m^2+i\varepsilon)m^2} \quad . \tag{5.14}$$

Thus we may look forward in the near future (when experimental results for $\rho(m^2)$ will be more complete) to being able to make a (first order in e^2) correction in QED calculations of the effects of loops involving hadrons.

In a recent measurement of $\sigma(e^+e^- \to \mu^+\mu^-)$ an effect of the contribution of the ϕ resonance to $\rho(m^2)$ has been observed. This comes from the interference of the graphs

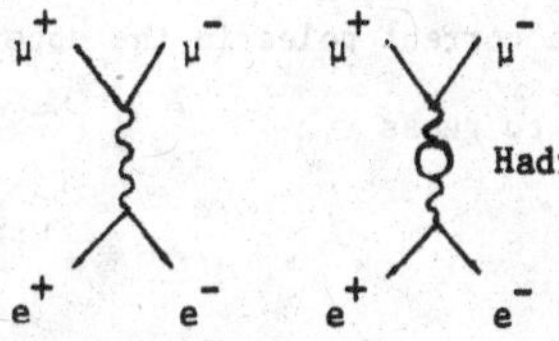

at a q^2 near m_ϕ^2. In the first order in which the effect contributes only the real part of $i\left(v(q^2)-v(0)\right)$ is seen, this is observed as a slight oscillation in σ near $q^2 = m_\phi^2$.

As an example we calculate the modification to $\sigma(e^+e^- \to \mu^+\mu^-)$ when "Hadrons" in the above graphs is the ϕ resonance only. In this case

$$\rho(m^2) = \frac{C}{(m^2-m_\phi^2)^2+\Gamma^2/4} \quad . \tag{5.15}$$

Substituting into (5.12)

$$i\left(v(q^2)-v(0)\right) = \frac{-q^2}{2\pi} \int_{4m_\pi^2}^{\infty} \frac{C}{\left((m^2-m_\phi^2)^2+\Gamma^2/4\right)} \frac{dm^2}{(m^2-q^2-i\varepsilon)m^2} \tag{5.16}$$

$$\mathrm{Im}\left(i\left(v(q^2) - v(0)\right)\right) = \frac{1}{2} \frac{C}{(q^2-m_\phi^2)^2+\Gamma^2/4} \tag{5.17}$$

$$\mathrm{Re}\left(i\left(v(q^2) - v(0)\right)\right) = \frac{-q^2}{2\pi} P \int_{-\infty}^{\infty} \frac{C}{(m^2-m_\phi^2)^2+\Gamma^2/4} \frac{dm^2}{(m^2-q^2)m^2} \tag{5.18}$$

where P means principal part.

The lower limit in the integral (5.18) can be set to $-\infty$ with negligible error because the resonance is very narrow. Carrying out the integral we find:

$$\mathrm{Re}\left(i\left(v(q^2) - v(0)\right)\right) = \frac{q^2C}{\Gamma}\,\frac{m_\phi^2(m_\phi^2-q^2)-\Gamma^2/4}{(m_\phi^4+\Gamma^2/4)\left((m_\phi^2-q^2)^2+\Gamma^2/4\right)} \quad . \tag{5.19}$$

The cross section for $\mu^+\mu^-$ production from e^+e^- is therefore

$$\sigma = \sigma_o\left[1 - \frac{4\pi e^2q^2C}{\Gamma}\,\frac{m_\phi^2(m_\phi^2-q^2)-\Gamma^2/4}{(m_\phi^4+\Gamma^2/4)\left((m_\phi^2-q^2)^2+\Gamma^2/4\right)}\right] \tag{5.20}$$

where σ_o is the cross section for the process in the absence of ϕ-photon interactions.

We need not have evaluated the integral in (5.18) to get the real part of $iv(q^2)$. What the relation (5.15) says is that $iv(q^2)$ is an analytic function with no poles in the lower half plane and such that $\mathrm{Im}\left(iv(q^2)\right) = \frac{1}{2}\,p(m^2)$. To obtain $\mathrm{Re}\left(iv(q^2)\right)$ all we need to do is guess an analytic function whose imaginary part is $\frac{1}{2}\,p(m^2)$ and which has the correct poles in the upper half plane. In this case, the function is easy to guess

$$iv(q^2) = \frac{C/\Gamma}{(q^2-m_\phi^2)-i\Gamma/2} \tag{5.21}$$

therefore

$$\mathrm{Re}\left(iv(q^2)\right) = \frac{C(q^2-m_\phi^2)/\Gamma}{(q^2-m_\phi^2)^2+\Gamma^2/4} \tag{5.22}$$

Making the subtraction at $q^2 = 0$ we get (5.19).

<u>Note: Annoying Point</u>

Let us calculate other $\mu\nu$ components in (5.8). First tx so $q'_\mu q'_\nu - \delta_{\mu\nu}q'^2 = q'_o\,Q_x$.

Therefore get

$$\frac{q'_o\,Q_x\,\theta(q'_o)}{q_o-q'_o+i\varepsilon} + \frac{-q'_o\,Q_x\,\theta(q'_o)}{q_o+q'_o-i\varepsilon} = \frac{2q_o\,Q_x\cdot q'_o\,\theta(q'_o)}{q'^2_o-q_o^2-i\varepsilon}$$

so this is OK because the factor $q_o\,Q_x$ multiplies the same integral as in the tt case. Trouble comes however in the xx case for then the coefficient is $Q_x^2 - (q'^2_o-Q^2)$ for the q'^2_o does not change directly to q^2 as required, but we have instead an extra correction of $q'^2_o - q_o^2$ which cancels the denominator so we get

$$V_{xx}(q) = \frac{i}{2\pi}(Q_xQ_x - \delta_{xx}q^2)\int\frac{p(m^2)dm^2}{q^2-m^2+i\varepsilon} + \frac{i}{2\pi}\int p(m^2)dm^2$$

or $$V_{\mu\nu}(q) = \frac{i}{2\pi}(q_\mu q_\nu - \delta_{\mu\nu}q^2)\int\frac{p(m^2)dm^2}{q^2-m^2+i\varepsilon} + \frac{i}{2\pi}(\delta_{\mu\nu}-\delta_{\mu t}\delta_{\nu t})\cdot C + iS_{\mu\nu} \quad (5.23)$$

where C is a constant (infinite no doubt) = $\int p(m^2)dm^2$.

We could get rid of C by a seagull type term but we are confused because the time ordered product alone seems not to be relativistically invariant. Apparently $\langle 0|[J_\mu(1), J_\nu(2)]|0\rangle = 0$ outside light cone isn't enough. Perhaps some limit on the singular behavior when $t_1 = t_2$ near $\bar{x}_1 = \bar{x}_2$ is also required (for relativistic invariance) and is not satisfied here. It is satisfied for every other real problem. Hereafter we drop it supposing some term like $\int \bar{A}\cdot\bar{A}\,d^3\bar{x}$ is added to the Hamiltonian to straighten it out. It is called the trivial Schwinger term. In quantum electrodynamics the trivial Schwinger term is controlled by calculating by regulators, i.e., propagating electrons of mass m^2 minus a term with mass $m^2 + \Lambda^2$ so $p(m^2)$ need not be positive and $\int p(m^2)dm^2$ is taken zero.

Certain functions and their transforms are important. Propagator

$$I_+^m(\bar{x},t) = \int\frac{e^{-ip\cdot(t,\bar{x})}}{p^2-m^2+i\varepsilon}\frac{d^4p}{(2\pi)^4} = -\frac{1}{4\pi}\delta(s^2) + \frac{m}{8\pi S}H_1^{(2)}(ms) \quad (5.24)$$

$$H_1^{\,2} = J_1 - iN_1$$

where $s = +\sqrt{t^2-x^2}$ in time like regions and $s = -i\sqrt{t^2-x^2}$ in space like regions. For large s, H is asymptotically $H \sim e^{-ims}$. For small s

$$I_+^m(\bar{x},t) = -\frac{1}{4\pi}\delta(s^2) + \frac{i}{4\pi^2 s^2} + \frac{m^2}{16\pi} - \frac{im^2}{8\pi^2}\left[\ell n(\tfrac{1}{2}ms) + \gamma - \tfrac{1}{2}\right] + \quad . \quad (5.25)$$

$\gamma = .5772\ldots$.

For free fields of mass m it is just $\langle 0|\{\phi(2), \phi(1)\}_T|0\rangle$.

<u>Commutator</u>.

The commutator function

$$C(\bar{x},t) = \text{sgn}\ t\cdot \text{Re}\left(I_+(\bar{x},t)\right) \quad (5.26)$$

$$\text{F.T. } C(\bar{x},t) = \text{sgn}\ q_0\ \delta(q^2-m^2) \quad . \quad (5.27)$$

For free fields it is $<0|\left[\phi(2),\ \phi(1)\right]|0>$

$$C(\bar{x},t) = \text{sgn}\ t\left[-\frac{\delta(s^2)}{4\pi} + \frac{m}{8\pi s} J_1(ms)\right]$$

for real s only. Outside the light cone it is zero. For small s

$$C(\bar{x},t) = \text{sgn}\ t\left[-\frac{\delta(s^2)}{4\pi} + \frac{m^2}{16\pi} - \frac{m^4 s^2}{128\pi} + \ldots\right] \quad . \tag{5.28}$$

Problem 5.1.

Show that the total cross section to produce any hadrons from $e^+ + e^-$ at energy E + E is

$$\sigma = \frac{(4\pi e^2)^2}{2q^2}\ p(q^2) \qquad \text{where } q^2 = 4E^2.$$

Problem 5.2.

Find $p(q^2)$ for muon pair production.

$$p(q^2) = \frac{1}{6\pi}\ \frac{q^2+2\mu^2}{q^2}\ \left(\frac{q^2-4\mu^2}{q^2}\right)^{1/2} \qquad \text{for } q^2 > 4\mu^2 \quad .$$

Problem 5.3.

Use $p(q^2)$ of problem 5.2 to find $v(q^2) - v(0)$ for the vacuum polarization of muons.

Low Energy Photon Reactions

Lecture 6

Pion Photoproduction Low Energy (0 to 2 GeV).

Reference: R. L. Walker, Phys. Rev. 182, 1729 (1969).

As our first experimental subject we take up low energy photoproduction of pions from nucleons. Among the individual reactions are

$\gamma + p \to \pi^+ + n$

$\gamma + p \to \pi^0 + p$

$\gamma + n \to \pi^- + p$

$\gamma + n \to \pi^0 + n$

The first two are the most extensively studied. Today data on various polarization and spin cases are also available. Evidently we are measuring things which depend on $<N\pi|J_\mu(q)|N>$ for $q^2 = 0$. The (total energy in the C.M.)^2 is often called s. $s = (p+q)^2 = m^2 + 2m\nu_{LAB}$ where ν_{LAB} is the energy of the photon in the lab system (for us ν goes from 0 to 2 say). These are not the only matrix elements from N states at these energies

$\gamma + p \to n + \pi^+ + \pi^0$ or $\gamma + p \to n^+ + \Lambda$, $\gamma + p \to p + n^0$

etc. are also possible but we discuss these later.

The most characteristic feature of the behavior is rapid changes in angular distributions with energy. This feature is characteristic of hadron collisions such as $\pi^- + p \to \pi^- + p$ also and is interpreted as due to resonances. The same resonances appear here, of course, as in π scattering. The photon data has been explained and analyzed in considerable detail on this basis. First we discuss the general "theory" of resonances.

A general element of the T matrix must have a special property of separability defined by the following circumstances. Suppose we imagine a collision like A + B + C → D + E + F. One possibility is that A + B hit first to make D + X in one part of space (via element $\langle DX|S|AB\rangle$ and then X as a real particle goes off across the room to hit C to make E + F via C + X → E + F. That is $\langle EFD|T|ABC\rangle$ must have an infinity--a singularity when the momenta satisfy $p_A + p_B - p_C = p_X$ and p_X^2 has the value M_X^2 for a real particle X. The residue of the singularity is $\langle EF|T|CX\rangle \cdot \langle XD|T|AB\rangle$. When p_X^2 is very near M_X^2 (and E_X is positive) the behavior is very much as if an ideal free particle of mass M_X^2 propagated so the amplitude varies as $(p_X^2 - M_X^2 + i\varepsilon)^{-1}$.

$$\langle DEF|T|ABC\rangle \approx \langle EF|T|CX\rangle \frac{1}{(p_A+p_B-p_C)^2 - M_X^2 + i\varepsilon} \langle DX|T|AB\rangle \quad . \qquad (6.1)$$

More specifically if S is written $S_{fi} = \delta_{fi} - (2\pi)^4\, i\delta^4(p_f - p_i)\, T_{fi}$ near a resonance (don't confuse T with T[A] used previously).

$$\langle CD|T_F|AB\rangle = \langle CD|T_F|Res\rangle \frac{1}{p^2 - (m_R^2 - i\Gamma_R m_R)} \langle Res|T|AB\rangle$$

where $(p_A + p_B - p_C)^2$ is sufficiently close to M_X^2 (and the sign of the energy is right). There must be a corresponding singular behavior for every way that the momenta can be combined into the momentum of some stable particle. The exact form of the singularity comes from an analytic extrapolation from the imaginary part

$$i\pi \langle EF|T|DX\rangle\langle XC|T|AB\rangle\, \delta(p_X^2 - M_X^2)$$

which unitarity demands because processes A + B → C + X and D + X → E + F are possible.

If a small perturbation (such as weak interaction) acts to make state X slightly unstable and of decay rate Γ_X it can be shown that the modification near the resonance is to replace

$$\frac{1}{p_X^2 - M_X^2 + i\varepsilon} \quad \text{by} \quad \left(p_X^2 - (m_X^2 - i\Gamma_X m_X)\right)^{-1}.$$

(There are also slight modifications of m_X and the residues, but the form is the same). The non-relativistic form is

$$\frac{1}{E - E_X - i\Gamma_X/2} \; .$$

In non-relativistic theory this is the form no matter what Γ is due to.

It has been found that such resonant behavior of the T matrix is exhibited by strong interactions. (Here the width Γ is not produced by weak external perturbation, but is a consequence of the strong interaction itself.) They have the following characteristics.

T varies in Breit Wigner fashion $\left(p^2 - (m_R^2 - i\Gamma_R m_R)\right)^{-1}$.

This only means something precise if Γ is small enough that the variations in the residue factors $\langle cd|T|R\rangle\langle R|T|ab\rangle$ are either negligible or well known (as near threshold behavior--there a proper variation of Γ with Q may have to be included).

The same resonances appear in many reactions at the same mass and width (given I-spin, angular momentum, strangeness, parity). The Table of Particle Properties is a table of such resonances. What is the meaning of the existence of resonances in strong interactions? There are examples of narrow resonances in other fields.

a) Atomic systems. The excited levels of atoms are stable except for coupling with light. The narrowness of the observed resonances in the emission spectrum of an atom, for example, is known to be due to the small value of the coupling e^2.

b) Nuclei. The coupling is known to be large but narrow resonances are observed. In this case the origin of the narrowness of the resonances can still be understood. In the decay of an excited nucleus of high A by the emission of a neutron; for example, it is hard for the wave function to concentrate enough energy on a single neutron to escape. Other effects such as centrifugal barriers account for the narrowness of resonances in nuclear physics.

None of the above effects seem to appear in particle physics. But also resonances are generally not so narrow, the width is about a fourth of the spacing to the next resonance in the same channel. The meaning of the appearanc of resonances in strong interactions is not understood. At first theoretical physicists did not expect resonances from strong coupling field theory. Then they realized if it were strong enough there were "isobaric states". But the exact meaning of the existence of these resonances for underlying theory is not clear.

Lecture 7

We discuss a number of problems in general in using resonances, in the specific form used by Walker.

a) How much is resonances and how much is background? Can the background below a resonance be simply tails of other resonances? If we write a resonance

$$\frac{a}{E-E_R-i\Gamma},$$

off resonance it is

$$\frac{a}{E-E_R}.$$

If a varies like $a(E) \approx a(E_R) + \alpha(E-E_R)$, then our resonance is

$$\frac{a(E_R)}{E-E_R} + \alpha .$$

Thus a general background α cannot be defined unless one is specific about the variation of <T> off resonance. Some people like to define <T> as pure constant and ask then if background is zero--but it may depend on how we write <T>, e.g. as γ_μ or other explicit forms and seems a bit arbitrary although it has merit of being precise.

In practice sometimes we know Γ must vary because we are close to a threshold where Γ varies as $Q^{2\ell+1}$ for small Q where ℓ is the orbital angular momentum. This form cannot be correct for large Q because it blows up too fast. To deal with that empirically Walker chose

$$\Gamma = \Gamma_0 \left(\frac{Q}{Q_R}\right)^{2\ell+1} \left(\frac{x^2+Q_R^2}{x^2+Q^2}\right)^{\ell}$$

for the rate a meson couples to a proton with angular momentum ℓ. For a photon a similar

$$\Gamma_\gamma = \Gamma_0 \left(\frac{k}{k_R}\right)^{2j} \left(\frac{x^2+k_R^2}{x^2+k^2}\right)^{j}$$

term was used. x was arbitrary (at about .350 GeV for all resonances except the 1236 where .160 was used). But changing x is like changing a(E),--it distorts the form and leaves the question--How big is the tail of a resonance? Impossible to answer except arbitrarily.

Problem.

The use of $\Gamma = \Gamma_0\ (Q/Q_0)^{1/2}$ has implications for the expression below threshold. These are not used by Walker (violating dispersion relations). What are they?

Theorists like to choose the infinitely narrow resonance approximation; a is constant, the background is

$$\frac{\text{a(at resonance)}}{E-E_R}$$

in defining resonance backgrounds--or rather "effects of resonances far from their resonant energy". This last concept in quotes is very subtle and hard to define but is used glibly for all kinds of things today in theory, as we shall see.

b) Can T be expressed entirely as a sum of resonant terms? Consider the case a + b → c + d

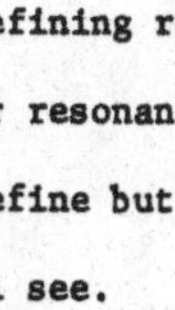

$$s = (p_a + p_b)^2$$
$$t = (p_c - p_a)^2$$
$$u = (p_a - p_d)^2$$

These are resonances in the s channel of the kind

$$\frac{1}{s-M_i^2(s)}$$

and have a sum

$$\Sigma \frac{C_i(t)}{s-M_i^{\,2}(s)} \quad . \tag{7.1}$$

Then there are resonances in the t channel from diagrams of the form

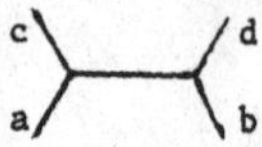

$$\Sigma \frac{D_i(s)}{t-M_i^{\,2}(t)} \quad . \tag{7.2}$$

For a real experiment none of these are resonant since $t < 0$, all are far away. A theoretical question is: Should we add (7.1) to (7.2) (and u resonances)? This only means something if (7.1) is precisely defined.

It is a tenet of theory (called Veneziano or extended duality) today that we should not add (7.1) and (7.2) but that in the sum of s channel resonances alone is the t channel behavior completely defined; this is supposed to apply to the imaginary part only.

An expression has been found for (7.1) which when summed gives something that can be written as a sum of t channel resonances (Veneziano). These are difficulties of definition in practice. What Walker did was to take a term for each resonance that he knew about, plus a background. In fact he wrote the amplitude as the sum of three parts:

(1) s-channel resonances
(2) the pion exchange
(3) background.

(1) The resonant masses and widths were taken from the πN scattering data which had been analyzed by amplitudes for each channel (of angular momentum, I spin, parity). A constant, the matrix $\langle R|J_\mu|N\rangle$ i.e. $\langle R|T|\gamma N\rangle$ for each resonance had to be determined empiracally, by adjusting to fit data. (The other factor $\langle \pi N|T|R\rangle$ was available from π scattering).

(3) The "background" was a slowly varying amplitude in each channel. Goodness of fit of parameters in (1) is judged if the background can be made to vary slowly, or at least smoothly. It was hoped that all these background terms would be real but some small imaginary was ultimately needed in some channels, possibly because resonances are left out, or have

inaccurate parameters. (Next time Walker does this he will relax the condition that the background amplitude be real.)

Lecture 8

We continue with the discussion of the terms used by Walker.

(2) t channel π pole. This is from diagram A

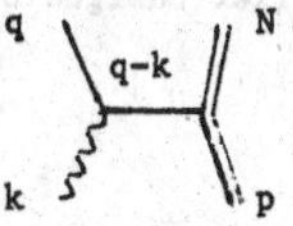

A

(a diagram which does not exist for π^0--but we will say more about this later).

It leads to a term like

$$(2q-k)\cdot e \frac{\sqrt{4\pi}\, e_\pi}{(q-k)^2-m_\pi^2} <\bar{u}_2|\gamma_5|u_1>\cdot\sqrt{2}\sqrt{4\pi g^2}$$

where $g<\bar{u}_2|\gamma_5|u_1>$ is an empirical determination of how π's couple at their pole, empirically $g^2 = 14.8$. This factor $t-m_\pi^2$ varies rapidly with t for small t, leading to a rapid angular dependence. Small t means an effect far away from the nucleon. The effect is that a nucleon has a chance to be seen as a nucleon with a π^+ around it spread relatively far out, as

$$e^{-m_\pi r}/r$$

in amplitude. If this virtual π is hit with a photon it can be sent out forwards and appear real ("unbound") having received its necessary energy and momentum from the photon. It goes fairly sharply forward--at higher energies at least and affects the amplitudes in all values of s-channel angular momentum including high values, where for low energies and momenta no resonances are large. So it gives the major contribution at the higher angular momenta. Since there the rapid variation with t near t = 0 is essential we are near the pole $m_\pi^2 = .02\ \text{GeV}^2$ and we can consider it trustworthy and accurate. To be sure, at the higher energies lower angular momenta involve it for t large and so the form $1/t-m_\pi^2$ etc. may be wrong or ambiguous (we mean we want to follow the principle--the contribution of poles is well defined only for parameters near their resonance). But here the amplitude is slowly energy dependent and its exact value for the lowest one or two angular momenta (s, p waves) is lost in the background amplitudes we are adding anyway. (No background was added for very high angular momentum).

The expression we have written for the amplitude is not gauge invariant, however. The π^+ springs from a source and doesn't conserve charge. It must be combined to other terms to make it completely gauge invariant. One obvious contender is the s channel resonance corresponding to the unexcited nucleon at 938. We can put these together as follows (let charges on initial nucleon be e_1, final nucleon e_2, pion e_π)

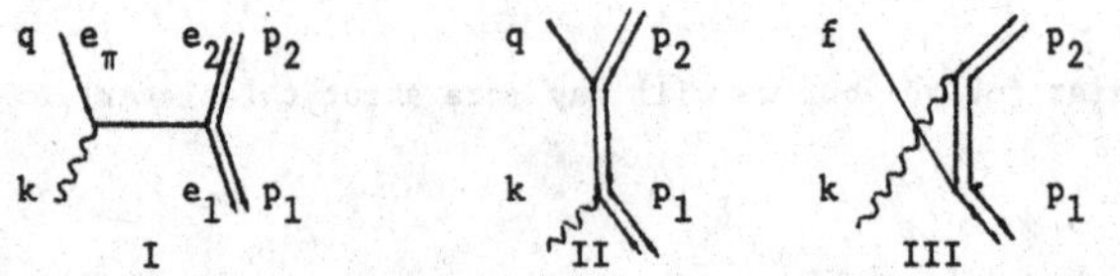

these give

$$M_I = 4\pi(2q-k)\cdot e \frac{1}{-2k\cdot q+k^2} (\bar{u}_2\gamma_5 u_1) e_\pi \sqrt{2} g \tag{8.1}$$

$$M_{II} = 4\pi\bar{u}_2\gamma_5 \frac{\not{p}_1+\not{k}+M}{k^2+2p_1\cdot k} \left(e_1\not{e} + \frac{\mu_1}{4M}(\not{e}\not{k} - \not{k}\not{e})\right) \bar{u}_1 \sqrt{2} g \tag{8.2}$$

$$M_{III} = 4\pi\bar{u}_2 \left(e_2\not{e} + \frac{\mu_2}{4M}(\not{e}\not{k} - \not{k}\not{e})\right) \frac{\not{p}_2-\not{k}+m}{k^2-2p_2\cdot k} \gamma_5 u_1 \sqrt{2} g \tag{8.3}$$

With $\not{e}$ replaced by $\not{k}$ the three expressions give

$$M_I = -4\pi e_\pi \sqrt{2} g\, (\bar{u}_2\gamma_5 u_1)$$

$$M_{II} = -4\pi e_1 \sqrt{2} g\, (\bar{u}_2\gamma_5 u_1)$$

$$M_{III} = -4\pi e_2 \sqrt{2} g\, (\bar{u}_2\gamma_5 u_1)$$

$$M_I + M_{II} + M_{III} = 0 \qquad \text{since } e_\pi = e_1 - e_2$$

<u>Note</u>

The inclusion of diagram III for the nucleon resonance reminds us of the question of whether we should not also have added to our low energy s channel resonances, such as the Δ, such diagrams as

The answer is certainly yes. They were not included by Walker but again they are relatively slowly varying for a positive energy ω. They are far off from their resonance (which appears below threshold at negative ω.)

The gauge invariance of $M_I + M_{II} + M_{III}$ is seen to hold even if the anomalous moment terms are omitted. These terms contribute only low angular momentum at low energy as does the entire II and III, and does not vary particularly rapidly with k. For small k all three terms have the expected 1/k singular behavior. The anomalous magnetic terms in II and III go to a constant as $k \to 0$ because there is a k in the numerator also. Since the background term was arbitrary except it was assumed that it was a) small in higher angular momentum states, b) slowly varying in energy; we could omit the anomalous moment terms and include them in the background. Walker did this and found, perhaps surprisingly, that if he left these anomalous moment terms out, the background was smaller than if he left them in. Thus his explicit definition of the "pion exchange term" is I + II + III with $\mu_1 = \mu_2 = 0$ (an expression he calls the "electric Born term").

We see that he could just have well left out the nonanomalous moment terms too and used a simpler expression--readily generalized to make gauge invariant in a rational way <u>any</u> pion exchange term. For

$$(\not p_1 + \not k + M)\not e = (2p_1 + k)\cdot e + \frac{1}{2}(\not k\not e - \not e\not k) - \not e(\not k_1 - M)$$

the last term vanishes because $\not p_1 = M$ on the initial state. The second term is like a magnetic moment term in that it has no pole as $k \to 0$; it is of low angular momentum and slow energy behavior; it can be taken into the background. Leaving out all such moment terms, as they only contribute to undefined background we find the pion exchange term can be written in a gauge invariant form:

"pion exchange term" =

$$4\pi\left(-e_\pi\frac{(2q-k)\cdot e}{(2q-k)\cdot k} + e_1\frac{(2p_1+k)\cdot e}{(2p_1+k)\cdot k} - e_2\frac{(2p_2-k)\cdot e}{(2p_2-k)\cdot k}\right)\sqrt{2}g\,(\bar{u}_2\gamma_5 u_1) \qquad (8.4)$$

(for π_o, $e_\pi = 0$ and $\sqrt{2}g$ is replaced by g, $e_1 = e_2$ so we still have a gauge invariant expression).

It is evident that the "pion exchange term" has no precise definition (some part of the nucleon resonance term is added). The ambiguity of exactly what part is academic because the ambiguity is slowly varying and easily absorbed in the definition of "background". We would suggest for a general case, at any k^2 (not only $k^2 = 0$) and any initial and final state that we take just the term

$$\langle f|\pi|i\rangle\ 4\pi\left(-e_\pi \frac{(2q-k)\cdot e}{(2q-k)\cdot k} + e_i \frac{(2p_i+k)\cdot e}{(2p_i+k)\cdot k} - e_f \frac{(2p_f-k)\cdot e}{(2p_f-k)\cdot k}\right) \qquad (8.5)$$

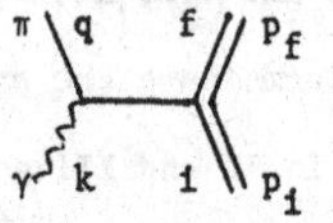

$\langle f|\pi|i\rangle$ is the amplitude for the diagram

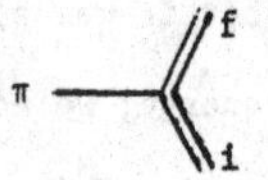

Since strictly this is off shell some definition by theory may be used, although all we can be sure of is that the residue on shell is correct. The "π^o exchange term" does not have $1/(t-m_\pi^2)$. The angular dependence is slow and shows no forward peak.

Any form like

$$\frac{(2q-k)\cdot e}{(2q-k)\cdot k} - \frac{a\cdot e}{a\cdot k}$$

for any a would be all right, but (8.5) has the virtue that the singular behavior as $k \to 0$ is physically correct.

I have gone through this long discussion to show that Walker committed no theoretical error of principle in his method of fitting. His adjustable parameters of fit were the desired quantities $\langle Res|J_\mu|N\rangle$ and the background amplitudes, adjusted to be as small and as slowly varying as possible. In this way all the data is distilled to produce theoretically significant numbers $\langle Res|J_\mu|N\rangle$. Walker gives the value of the matrix element as A:

$$A = \sqrt{4\pi e^2}\ (2M2M_2 2Q)^{-1/2}\ \langle f|J\cdot e_{pol}|i\rangle$$

rather than $\langle f|J\cdot e|i\rangle$ directly. Suppose a resonance of mass m_1 decays into a nucleon energy E_2, and a photon each of momentum Q. $(Q = (m_1^2-m_2^2)/2m_1)$. If the photon goes off in the z direction with positive helicity +1, there are two helicities possible for the proton -1/2, +1/2, in which case the helicity of the decaying resonance is +3/2 or +1/2. A is given for each case, and for whether it is proton or neutron (see Table 8.1).

Quark Model of Resonances

The Quark Model

The various baryon resonances have been successfully classified by the quark model; they are as if they were made of three quarks. Each quark has spin 1/2; three possible unitary spin states u, d, s; u, d are I spin +1/2, -1/2 strangeness zero; s is I spin 0, s = -1; charges are +2/3 for u, -1/3 for s, d. It is assumed that the state is separable in its dependence on unitary spin, regular spin, and some additional internal coordinate or coordinates having characteristics of internal orbital motion.

The nucleon octet and the lowest decimet require no "orbit" contribution, but the next set of levels of odd parity are understood as getting their parity and some of their angular momentum from a first excitation of an orbital angular momentum. These "quarks" can be looked upon as abstract indices to describe a wave function (in which case we need three indices each of which can be a unitary and a spin containing index, taking therefore 3 x 2 (or 3 x 4 using Dirac spinors?) values and another 4-space vector index for the angular momentum. Alternatively if the picture is taken more literally as particles going around each other we would be lead to expect two internal degrees of freedom and two angular momenta which could combine. This leads to the expectation of a greater variety of states than does the other view, but this excess only arises at second internal excitation, or states of positive parity of higher masses. Today there is no sure evidence that these two angular momenta are necessary (a state needing them on the 1970 table is now considered uncertain). However, we will suppose they exist, and will study a rather "real" physical view of the model. The abstract theory with one angular momentum has not been formulated clearly enough to do calculations with it--it may not exist as a true alternative.

I will just briefly remind you of the ideas and elementary consequences of the quark model. A quantum state of three objects can have one of 4 possible properties under permutations of these objects. It may be symmetric S, or change sign, antisymmetric A, or in one of two states of mixed symmetry α, β which upon permutations become linear combinations of themselves. Thus if X represents the two mixed states, combine S with X say makes a state of mixed symmetry X, A with A is symmetric S, X with X makes with different linear combinations of the four states either S, X or A, thus

product symmetries	S	X	A
S	S	X	A
X	X	S,X,A	X
A	A	X	S

With unitary spin there are three choices for each quark or 27 states. There are 10 symmetric states $|uuu>_s$, $|uud>_s$ etc. and only 1 antisymmetric $|uds>_A$ (since each must be different). The remaining 16 are 8α and 8β. Thus the unitary symmetry is

S = $\underset{\sim}{10}$ decimet

X = $\underset{\sim}{8}$ octet

A = $\underset{\sim}{1}$ singlet .

Only such SU_3 multiplets are found among the baryons. For spin with two possibilities there is no antisymmetric state. The symmetric is spin 3/2 quartet, the X (two) is spin 1/2 doublet, symbolized by superscript 4 or 2.

Therefore combining spin and unitary spin we can make the following multiplets (SU_6 multiplets). $\mathcal{S}$ can come from S·S or X·X via chart above so

$|S\rangle = {}^4\underset{\sim}{10}$ or ${}^2\underset{\sim}{8}$ 56 states
$|X\rangle = {}^2\underset{\sim}{1}\ {}^2\underset{\sim}{8}\ {}^4\underset{\sim}{8}$ or ${}^2\underset{\sim}{10}$ 70 states
$|A\rangle = {}^4\underset{\sim}{1}$ and ${}^2\underset{\sim}{8}$ 20 states.

Finally we combine this with internal orbital motion. We can think of a harmonic oscillator force for definiteness. The answer depends on the degree of excitation of the orbits N.

We are to make an overall completely symmetrical state.

N = 0. No internal motion. Therefore pure $|S\rangle = \underline{56} = {}^4\underset{\sim}{10}_{3/2},\ {}^2\underset{\sim}{8}_{1/2}$ the well known lowest decimet and octet.

N = 1. One unit of excitation has the X symmetry. Because the CM cannot move (relative coordinates only) $X_1 + X_2 + X_3$ where X is a position of a quark is impossible. Only differences can be involved, the $|\alpha\rangle$, $|\beta\rangle$ states:

$$|\alpha\rangle \quad \frac{1}{\sqrt{6}}(2X_1 - X_2 - X_3)$$

$$|\beta\rangle \quad \frac{1}{\sqrt{2}}(X_2 - X_3) \quad .$$

To get overall $|S\rangle$, this orbital $|X\rangle$ must be combined with unitary spin and spin $|X\rangle$ or 70. Finally when the unit of angular momentum is combined with the spin in various ways to make different total angular momentum j, all of odd parity, we find

${}^2\underset{\sim}{1}_{1/2}$ ${}^2\underset{\sim}{1}_{3/2}$

${}^2\underset{\sim}{8}_{1/2}$ ${}^2\underset{\sim}{8}_{3/2}$

${}^4\underset{\sim}{8}_{1/2}$ ${}^4\underset{\sim}{8}_{3/2}$ ${}^4\underset{\sim}{8}_{5/2}$

${}^2\underset{\sim}{10}_{1/2}$ ${}^2\underset{\sim}{10}_{3/2}$

All these multiplets (except the ${}^4\underset{\sim}{8}_{3/2}$) have been found.

Note

If an object can be in one of a number of conditions x, y, z... we can with three such objects form states like $|zxy\rangle$ meaning the first object is in condition z, the second x, third y. From this and its permutations $|xzy\rangle$ etc.

we can form the states of definite symmetry

$$|S\rangle = |xyz\rangle_S = \frac{1}{\sqrt{6}}(|xyz\rangle + |xzy\rangle + |yxz\rangle + |yzx\rangle + |zxy\rangle + |zyx\rangle)$$

$$|\alpha\rangle = |xyz\rangle = \frac{1}{2\sqrt{3}}(|xyz\rangle + |xzy\rangle + |yxz\rangle + |yzx\rangle - 2\,|zxy\rangle - 2\,|zyx\rangle)$$

$$|\beta\rangle = |xyz\rangle = \frac{1}{2}(|xyz\rangle - |xzy\rangle - |yzx\rangle + |yxz\rangle)$$

$$|A\rangle = |xyz\rangle_A = \frac{1}{\sqrt{6}}(-|xyz\rangle + |xzy\rangle - |yzx\rangle + |yxz\rangle - |zxy\rangle + |zyx\rangle) \quad .$$

If two say xy are the same state $y = x$ replace $|xyz\rangle + |yxz\rangle$ by $\sqrt{2}|xxz\rangle$. If three are equal only the S survives $|xxx\rangle_S = |xxx\rangle$.

Combining two such states in a product we find

$$S\cdot S = S \qquad A\cdot S = A \qquad S = \frac{1}{\sqrt{2}}(\alpha\alpha + \beta\beta)$$

$$S\cdot\alpha = \alpha \qquad A\cdot\alpha = \beta \qquad \alpha = \frac{1}{\sqrt{2}}(-\alpha\alpha + \beta\beta)$$

$$S\cdot\beta = \beta \qquad A\cdot\beta = -\alpha \qquad \beta = \frac{1}{\sqrt{2}}(\alpha\beta + \beta\alpha)$$

$$S\cdot A = A \qquad A\cdot A = S \qquad A = \frac{1}{\sqrt{2}}(-\alpha\beta + \beta\alpha).$$

With these rules and the rules for combining angular momenta any specific state in the quark model can be constructed.

TABLE 8.1 Photoelectric matrix elements.

State	Multiplet	J_z	I_z	$\langle f \vert J_\mu e_\mu \vert i \rangle / F$	$A(\text{GeV})^{-1/2}$	$A^{NR}(\text{GeV})^{-1/2}$	$A^{exp}(\text{GeV})^{-1/2}$
$P_{33}(1236)$	$^4\underset{\sim}{10}_{3/2}[\underset{\sim}{56},0^+]_0$	+3/2	p	$-\sqrt{6}\,\rho$	−0.190	−0.178	−0.244
		+1/2	p	$-\sqrt{2}\,\rho$	−0.110	−0.103	−0.136
$D_{13}(1520)$	$^2\underset{\sim}{8}_{3/2}[\underset{\sim}{70},1^-]_1$	+3/2	p	$+\sqrt{\Omega}$	+0.115	+0.112	+0.151
		+1/2	p	$-\sqrt{3}\,\lambda\rho + \sqrt{\frac{1}{3}}\sqrt{\Omega}$	−0.036	−0.029	−0.026
		+3/2	n	$-\sqrt{\Omega}$	−0.115	−0.112	−0.132
		+1/2	n	$+\sqrt{\frac{1}{3}}\,\lambda\rho - \sqrt{\frac{1}{3}}\sqrt{\Omega}$	−0.033	−0.030	
$S_{11}(1535)$	$^2\underset{\sim}{8}_{1/2}[\underset{\sim}{70},1^-]_1$	+1/2	p	$+\sqrt{\frac{3}{2}}\,\lambda\rho + \sqrt{\frac{2}{3}}\sqrt{\Omega}$	+0.165	+0.160	+0.096
		+1/2	n	$-\sqrt{\frac{1}{6}}\,\lambda\rho - \sqrt{\frac{2}{3}}\sqrt{\Omega}$	−0.115	−0.109	−0.118
$D_{15}(1670)$	$^4\underset{\sim}{8}_{5/2}[\underset{\sim}{70},1^-]_1$	+3/2	p	0	0	0	0.040 †
		+1/2	p	0	0	0	∼ 0
		+3/2	n	$-\sqrt{\frac{3}{5}}\,\lambda\rho$	−0.057	−0.053	
		+1/2	n	$-\sqrt{\frac{3}{10}}\,\lambda\rho$	−0.041	−0.038	
$S_{31}(1650)$	$^2\underset{\sim}{10}_{1/2}[\underset{\sim}{70},1^-]_1$	+1/2	p	$-\sqrt{\frac{1}{6}}\,\lambda\rho + \sqrt{\frac{2}{3}}\sqrt{\Omega}$	+0.050	+0.047	

TABLE 8.1(cont.)

State	Multiplet	J_z	I_z	$\langle f \| J_\mu e_\mu \| i\rangle / F$	$A(\mathrm{GeV})^{-1/2}$	$A^{NR}(\mathrm{GeV})^{-1/2}$	$A^{exp}(\mathrm{GeV})^{-1/2}$
$D_{33}(1670)$	${}^2 10_{3/2}[70,1^-]_1$	+3/2	p	$+\sqrt{\Omega}$	+0.091	+0.091	
		+1/2	p	$+\sqrt{\frac{1}{3}}\,\lambda\rho+\sqrt{\frac{1}{3}}\sqrt{\Omega}$	+0.095	+0.092	
$P_{11}(1470)$	${}^2 8_{1/2}[56,0^+]_2$	+1/2	p	$(+\sqrt{\frac{3}{4}}\,\lambda\rho)\,\lambda$	+0.028	+0.032	
		+1/2	n	$(-\sqrt{\frac{1}{3}}\,\lambda\rho)\,\lambda$	-0.019	-0.020	
$F_{15}(1688)$	${}^2 8_{5/2}[56,2^+]_2$	+3/2	p	$(+\sqrt{\frac{4}{5}}\sqrt{\Omega})\,\lambda$	+0.064	+0.070	+0.139
		+1/2	p	$(-\sqrt{\frac{9}{10}}\,\lambda\rho+\sqrt{\frac{2}{5}}\sqrt{\Omega})\,\lambda$	-0.011	-0.015	~ 0
		+3/2	n	0	0	0	~ 0
		+1/2	n	$(+\sqrt{\frac{2}{5}}\,\lambda\rho)\,\lambda$	+0.038	+0.041	
$\omega(784)$	${}^3S_1$	0	π	$\sqrt{\frac{1}{2}}\,\rho$	0.297		0.22 ± 0.02
$\Lambda(1520)$	${}^2 1_{3/2}[70,1^-]$	+3/2	Λ	$+\frac{5}{6}\sqrt{\Omega}$	+0.109		} 0.097 ± 0.010
		+1/2	Λ	$-\frac{5}{6}\left[\sqrt{3}\,\lambda\rho-\sqrt{\frac{1}{3}}\sqrt{\Omega}\right]$	+0.012		

$$A = \sqrt{4\pi e^2}\ (2m_1 \cdot 2m_2 \cdot 2Q)^{-1/2}\ \langle f|J_\mu e_\mu|i\rangle$$

Resonance mass m_1 goes to photon momentum Q and final hadrons (see I_z column) mass m_2. J_z is the resonance spin projection along the direction of motion of the decay photon of helicity +1.

$$Q = \frac{m_1^2 - m_2^2}{2m_1}\ ,\ \lambda = \sqrt{\frac{2}{\Omega}}\, Q,\ \rho = \sqrt{\Omega}\ \frac{2m_1}{m_1+m_2}\ \lambda = \sqrt{2}\ (m_1 - m_2);$$

$$F = \exp\left(\frac{-m_1^2 Q^2}{\Omega(m_1^2+m_1^2)}\right) \approx 1$$

$$\Omega = 1.05\ (\text{GeV})^2$$

Lecture 9

The Quark Model (cont.)

N = 2. Two orbital excitations each of type X can make S, X, A and the total orbital angular mom. can be 2, 1, 0 as it turns out the state of symmetry A has total L = 1, states S or X can have 2 or 0. Hence we can make

$$[56,0^+]\ ;\ [70,0^+]\ \text{and}\ [20,1^+] = {}^4\underset{\sim}{1}_{1/2}\quad {}^4\underset{\sim}{1}_{3/2}\quad {}^4\underset{\sim}{1}_{5/2}$$

$$[56,2^+]\ ;\ [70,2^+] \qquad {}^2\underset{\sim}{8}_{1/2}\quad {}^2\underset{\sim}{8}_{3/2}.$$

The 20 has not been seen. It has no matrix element of any operator (like J_μ) operating on one quark at a time to the proton. Two quarks at least must have their motion changed to get to the 20 from the fundamental $[56,0^+]_{N=0}$.

We have definite need for the $[56,2^+]$ because of a Δ of spin $7/2^+$ which can be gotten in no other way. Two other Δ's of this set of decimets at $j = 1/2^+, 3/2^+$ have also been seen at roughly the same energy. This whole multiplet is expected as a Regge recurrence of the fundamental $[56,0^+]_{N=0}$ and is 2.1 $(\text{GeV})^2$ in mass^2 above. This also accounts well for the octet at $5/2^+$ ($m^2 = 2.85$).

A puzzling state is the $1/2^+$ Roper resonance at 2.16. It is best fitted as evidence of the $[56,0^+]_{N=2}$. What is puzzling about this state is its low mass as compared to the $^2 8_{5/2}$ at 2.85 which is best fitted in $\left[56,2^+\right]_{N=2}$. These two states are identical in their SU(6) properties. The quark model with harmonic dynamics predicts them to be of the same mass since they both correspond to two internal excitations. The large mass difference shows that harmonic dynamics may be too simple. The Roper resonance is a breathing mode oscillation of the nucleon; it is likely that all breathing modes of oscillation lead to a resonance of lower mass than those with a net orbital angular momentum. The corresponding decimet in the $\left[56,0^+\right]_{N=2}$ has not been seen.

These $\left[56,0^+\right]_{N=2}$, $\left[56,2^+\right]_{N=2}$ are the states that are expected if there were only one internal degree of freedom. The 70's would <u>not</u> be expected. How good is the evidence that they are there. There are three states that may require them:

$1/2^+ N(1780)$ possibly $[70,0^+]$ $m^2 = 3.16$

$3/2^+ N(1860)$ possibly $[70,2^+]$ $m^2 = 3.46$

$7/2^+ N(1990)$ $[70,2^+]$ $m^2 = 3.96$.

The latter state would have to be $\left[70,2^+\right]$; it would be impossible with one degree of freedom only--but it is not well established. (It was on the 1970 Particle tables, but not on the 1971 tables!) The first could also be $\left[56,0^+\right]_{N=4}$ a higher "breathing" mode excitation of the nucleon, the set being nucleon $m^2 = .881$, Roper $m^2 = 2.16$, $N(1780) m^2 = 3.16$.

The second, if only one internal motion is allowed, would have to be the missing octet from the $\left[56,2^+\right]$ at $J = 3/2$. If so its mass^2 = 3.46 is very different from the mass^2 = 2.85 of the $J = 5/2$ state so the usual rule of small spin orbit coupling would have to be abandoned.

Thus the evidence is in favor of requiring two internal degrees of freedom but it is not conclusive.

<u>Problem</u>.

A good research problem is to carefully study the properties of the 20 via the quark model and propose the wisest way to look for it experimentally.

The difficulty in observing the 20 is that if we assume we probe the baryons through an operator which acts on one quark at a time (regardless of the nature of the operator) then the 20 cannot be reached from the nucleon. The reason for this is that the 20 is antisymmetric in its SU_6 indices while the nucleon is totally symmetric. Operating on one of the quarks in the nucleon still leaves the other two in a symmetric state.

There are many states possible at N = 3. The most important is surely the Regge recurrence of $\left[70,1^-\right]$ at $\left[70,3^-\right]$. A few states of this are apparently known.

N = 4. A Δ of spin 1/2 probably from $\left[56,4^+\right]_{N=4}$ is known.

Calculation of Matrix Elements.

The simplest application of the quark model is to calculate matrix elements assuming a nonrelativistic Schrödinger equation for the quarks. There are ambiguities of factors like m_2/m_1 or E_2/m_2 (energies and masses of final and initial state) but for our photoelectric matrix elements they are small. See R. Walker "Single pion photoproduction in the resonance region" International Symposium on Electron and Photon Interactions at High Energies, Daresbury, England, 1969. See also references therein.

One simply works out the states of a system with Hamiltonian of three harmonic oscillators:

$$H = \frac{1}{2m}(\vec{p}_1^{\,2} + \vec{p}_2^{\,2} + \vec{p}_3^{\,2}) + \frac{m\omega^2}{2}\left((\vec{x}_1 - \vec{x}_2)^2 + (\vec{x}_2 - \vec{x}_3)^2 + (\vec{x}_1 - \vec{x}_3)^2\right). \tag{9.1}$$

One takes the spin and unitary spin dependences to factor out from the internal motion.

The interaction of one quark, say 1, with an electromagnetic field is taken to be

$$e_1\left[\left(\frac{g}{2m}\right)\vec{\sigma}_1\cdot\vec{B}(x_1) + \frac{1}{m}\,\vec{p}_1\cdot\vec{A}(x_1)\right]$$

where e_1 is the charge on the quark. Matrix elements of this operator (with A the appropriate plane wave for a photon) between states n of the Hamiltonian. (9,1) gives us our numbers for $\langle Res|J_\mu|p\rangle$ to compare to experiment.

The quantities m, ω, g, are parameters; g is the gyro-moment of the quark. The Dirac value g = 1 works very well and it was chosen to be 1. The value of ω is guessed as 400 MeV so the level spacing will be roughly right--the value of m is chosen (as 340 GeV) to fit the proton magnetic moment. The values of the current elements that result are given in table 8.1 next to last column

$A_{\text{non-rel.}} = A^{NR}$. It is seen that there is remarkably good agreement with the known values. All the signs are right and small values are predicted to be small, sometimes by selection rules generated by the model--sometimes by cancellation of terms for orbit and spin coupling. (The terms containing ρ in the column for $\langle f|J_\mu e_\mu|i\rangle$ are the ones contributed by spin--the formulas are from a relativistic modification. They are virtually the same as the nonrelativistic case). Finally the orders of magnitude are generally very close--the worst is the $F_{15}(1688)$ with +3/2 helicity on proton which is off by a factor 2. It is as possible that this is an error of the theory, as that it is a systematic error of Walker's fit where he omitted, for simplicity, a number of resonances which the quark model says should not be small.

Several questions arise if this is real. First a non-relativistic theory is surely not valid containing particles of mass 340 MeV in which the first excitation is 400 MeV. Second, with such low quark mass there should be $QQQQ\bar{Q}$ states.

We have used Bose statistics. If the quarks can be free it can be proven that spin 1/2 implies Fermi statistics. The only way out is to assume three kinds or "colors" of quarks for each SU_3 type and say that the bound states are pure singlets in the new index. For example, baryons must be in the 3Q state containing one of each color thus antisymmetric in the color, to say now that quarks obey Fermi statistics requires symmetry in the other indices. (A formal way of dealing with this called parastatistics can be shown to be exactly equivalent to the three colors of Fermi quark theory.)

Lecture 10

Feynman, Kislinger and Ravndal, Phys. Rev. (1971), tried to remedy the point that a nonrelativistic theory was ambiguous in how you applied it to a problem of relativistic kinematics by making a theory where the rules at least were relativistically stated. Such a theory to be complete would have to be a relativistic field theory which is too complicated to calculate. They made a simple theory of states, but at the expense of unitarity--which meant considerable trouble later on--but at least relativistically unambiguous rules could be formulated. The net result for our photon matrix elements is to give almost exactly the same numerical result as the non-relativistic case--so it is not of great interest here. (It does so with one less arbitrary constant, however.) Nevertheless we outline the method.

They noted that the harmonic oscillator Hamiltonian could be written $2mH = p^2 + m^2\omega^2x^2$ and that 2mH was nearly the difference of the squares of the energy m + W (or mass) of the states. Thus to get $(\text{mass})^2$ to be equally spaced they could use an eigen-vector operator for mass^2 which was a harmonic oscillator. The only constant is $m^2\omega^2$ which they call Ω^2, so the quark mass disappears on the right hand side. Thus let

$$K = 3(p_a{}^2 + p_b{}^2 + p_c{}^2) + \frac{\Omega^2}{36}\ (u_a - u_b)^2 + (u_b - u_c)^2 + (u_c - u_a)^2 + C \tag{10.1}$$

C is a constant, p_a is 4-dimensional momentum, u_a = 4-dimensional position of quark a. This in reciprocal is a propagator 1/K. Its poles (K = 0) give mass^2 as seen as follows:

Take out the cm. motion

$$\begin{aligned} p_a &= \frac{1}{3}\,P - \frac{1}{3}\,\xi \\ p_b &= \frac{1}{3}\,P + \frac{1}{6}\,\xi - \frac{1}{2\sqrt{3}}\,\xi \\ p_c &= \frac{1}{3}\,P + \frac{1}{6}\,\xi + \frac{1}{2\sqrt{3}}\,\xi \end{aligned} \tag{10.2}$$

to get relative momentum operators ξ, η (with coordinates x, y) find

$$\begin{aligned} K &= p^2 - N \\ -N &= \frac{1}{2}\ \xi^2 + \frac{1}{2}\ \eta^2 + \frac{\Omega^2}{2}\,x^2 + \frac{\Omega^2}{2}\,y^2 \end{aligned} \tag{10.3}$$

N is the Hamiltonian for two oscillators. Thus states of eigen-value $N = N_o$ have poles in K as $(p^2 - N)^{-1}$, or propagate as particles of $\text{mass}^2 = N_o$. Evidently these are separated by Ω. Regge slopes indicate mass^2 increases by $1.05\ (\text{GeV})^2$ per unit of angular momentum, hence take Ω = 1.05 GeV.

(For the baryon photoelectric matrix elements there are no further (effective) adjustable constants.)

If K is perturbed by δK the propagator becomes

$$\frac{1}{K} + \frac{1}{K}\,\delta K\,\frac{1}{K} + \frac{1}{K}\,\delta K\,\frac{1}{K}\,\delta K\,\frac{1}{K}$$

so in first order the disturbance δK becomes

$$\frac{1}{p^2-m_f^2}\,\langle f|\delta K|i\rangle\,\frac{1}{p^2-m_i^2}$$

or the appropriate perturbations of m^2--which is the relativistic T matrix elements defined by the usual relativistic rules--is just the matrix elements of δK between eigen states of N.

What about spin? The lack of much spin orbit coupling suggests that spin factors from the rest of the wave function--thus the operator K is the same expression times a wave function with three dirac 4 component indices although it does not affect those indices in no electromagnetic field. However p_a^2 is interpreted as its equal $(\not{p}_a)^2$ so when an electromagnetic field is operating the operator is

$$K_{inA} = 3\left[\left(\not{p}_a - e_a\not{A}_a(u_a)\right)\left(\not{p}_a - e_a\not{A}_a(u_a)\right) + \left(\not{p}_b - e_b\not{A}_b(u_b)\right)^2 + \left(\not{p}_c - e_c\not{A}(u_c)\right)^2\right] + \frac{\Omega^2}{36}\left[(u_a - u_b)^2 + (u_b - u_c)^2 + (u_c - u_a)^2\right] \tag{10.4}$$

e_a, e_b, e_c are the charges on the three quarks.

The first order perturbation thus gives the current operator for momentum q:

$$J_\mu^V = 3 \sum_{\alpha=a,b,c} e_\alpha\left(\not{p}_\alpha\gamma_\mu e^{iq\cdot u_a} + \gamma_\mu e^{iq\cdot u_a}\not{p}_\alpha\right). \tag{10.5}$$

Matrix elements of this are taken between states of the system.

But the system has too many states. (1) The 4-dimensional oscillator has time-component states which have negative norm (or negative energy - FKR prefer the former but see below). To avoid this it is assumed they are not excited - so the experimental states satisfy the subsidiary condition of being in the rest state of the oscillator in the direction of the 4-momentum P_μ of the state (if a_μ^* and b_μ^* are creation operators for the two oscillators),

$$P_\mu a_\mu^*|\phi\rangle = 0$$

$$P_\mu b_\mu^*|\phi\rangle = 0 \tag{10.6}$$

(2) By taking ϕ to involve Dirac spinors twice as many states as desired are included (Q and $\overline{Q}$). To avoid the parts of the spinor appropriate to antiquark states other subsidiary conditions are assumed.

$$P_\mu \gamma_{\mu a}|\phi\rangle = m|\phi\rangle$$

$$P_\mu \gamma_{\mu b}|\phi\rangle = m|\phi\rangle$$

$$P_\mu \gamma_{\mu c}|\phi\rangle = m|\phi\rangle \qquad (10.7)$$

where m is the mass of the state ($m^2 = P_\mu P_\mu$). (3) The current is not gauge invariant unless the mass^2 differences used are exactly NΩ (i.e. C is a true constant). These matrix elements are computed with the true experimental mass^2 which may not have exactly a separation NΩ. Transitions from ^{2}8p to quartet states are unaffected. Also the F_{15}(1688) is unaffected for its mass^2 is almost exactly 2Ω above the proton. The uncertainty involved for all elements of the table (except $\omega\to\pi\gamma$) is probably very small.

Omitting such states [10.6) and (10.7)] means a complete set is not used and, since the omitted states have negative norms, means that matrix elements will be generally too large by ever increasing factors as the mass discrepancy of initial and final state rises. FKR compute all kinds of meson decay widths (via PCAC using the divergence of the axial current for the pion coupling operator) and find this to be the case. To compensate they divide by a factor $(\bar{u}_2 u_1)^{-3}$ for the spins, and include a cutoff factor

$$\exp\left(-\frac{\alpha\, Q^2 m_1^2}{\Omega(m_1^2 + m_2^2)}\right) \qquad (10.8)$$

(with α determined empirically to fit, $\alpha \approx 1/\Omega$.

These do not seriously affect our matrix elements, of current, for lower states (the theoretical 1688 would be 40% higher without them) they have been included in the numbers given in table (8.1). It is seen that no essential modification of the non-relativistic model is implied for these matrix elements of J_μ.

The table also contains one meson entry, that for $\omega \to \pi + \gamma$; here relativity is very important. The agreement is good but recent experiments give still lower results (more like .15). The value comes from observation of the branching ratio rate of the ω into $\pi + \gamma$.

Beside these non-diagonal matrix elements there is, of course, the diagonal element $\langle \text{Proton} | J_\mu | \text{Proton} \rangle$. This is known for all momentum Q as the form factor of the proton - we discuss here only the low q results, the charge (which of course comes out right) and the magnetic moment. The quark model (as used here) is quite unsuccessful there. The relativistic theory gives $\mu = 3.00$ Nuclear magnetons for the proton and - 2.00 for the neutron (this compares to 2.79 and -1.91 experimentally). It (like SU_3) gives $\mu_\Lambda = 1/2\ \mu_N =$ -1.00, but in the magneton of the Λ hence -0.84 Nuclear magnetons. Experiment is -0.70 ± 0.07. (The reason it is magnetons of the mass of the state is instead of energy perturbation $= \mu(\sigma \cdot B)$ we have perturbation of mass^2 = $m\mu(\sigma \cdot B)$ so the values of $m\mu$ are the simple numbers).

Supposing that all this serves as evidence for the quark model, where do we go from here? We have these possibilities:

I Extensions and improvements in principle of theory:

a) Extension to other current matrix elements. Naturally axial current elements can be calculated directly from the same model (put $\not{A} \to \gamma_5 \not{A}$) and a few are known by β-decay. They come out wrong by the same factor, if theory is multiplied by .71 it agrees with experiment (e.g., for g_A we get 5/3 from the model and 1.23 experimentally). Using PCAC amplitudes for resonance pion emission to nucleon can be calculated with general success, except for a number of very sensitive matrix elements which come out as the difference of two large terms. For details see FKR - we will discuss the photoelectric matrix elements for $q^2 = 0$ later.

b) States of negative norm. These cause all kinds of ambiguities and uncertainties. First, nobody has any idea for the factors introduced by the spin (except to consider the wave equation to be that for a two-component spinor wave function - this has not been worked out - on the face of it, it appears it would violate parity.) Second, we have negative norm time-like states. In this regard, a suggestion was made by Fujimara, Kobayashi , Narniki, Progr. Theor. Phys. (Kyoto) <u>49</u> 193 (1970). FKR use the "wave function" $\exp[t^2 - x^2 - y^2 - z^2]$ for the ground state of the oscillators (all others are simply polynomials times this factor.) The exploding e^{t^2} can only be controlled by negative norm. FKN use instead $\exp[-t^2 - x^2 - y^2 - z^2]$ which they point out is also a solution of

$$\left(\frac{\partial^2}{\partial t^2} - \frac{\partial^2}{\partial x^2} - \frac{\partial^2}{\partial y^2} - \frac{\partial^2}{\partial z^2} - (t^2 - x^2 - y^2 - z^2)\right)\phi = E\phi \tag{10.9}$$

which is after all the sum of four independent oscillators for t, x, y, z each of which can be in its ground state. The time oscillator now contributes negatively to mass^2, and we can have states below the ground state and even states of negative m^2. Again we shall have to say something like "time states are not excited" or something - but at least it is an alternative view.

Thus the wave functions in momentum variables have the form

$$\exp\frac{1}{2\Omega}\left(p^2 - \frac{(P\cdot p)\ (P\cdot p)}{P^2}\right) \tag{10.10}$$

where P is the four-vector momentum of the overall state $P^2 = M_{Res}^{\ 2}$. (The ground state wave function is simply this factor - others multiply by polynomials.) Matrix elements of currents $e^{iq\cdot x}$ between two states P_1 to P_2; $P_2 - P_1 = q$ involve overlap integrals like

$$\int \exp\frac{1}{2\Omega}\left(p^2\frac{(P_1\cdot p)(P_1\cdot p)}{P_1^{\ 2}}\right)\delta^4(p-p'-q)\exp\frac{1}{2\Omega}\left(p'^2 - \frac{(P_2\cdot p')(P_2\cdot p')}{P_2^{\ 2}}\right)dp^4 dp'^4 \tag{10.11}$$

which is easily integrated to give a form factor

$$F = \frac{M_1^{\ 2}M_2^{\ 2}}{(P_1\cdot P_2)^2}\exp\frac{1}{\Omega}\left(q^2 - \frac{2(P_1\cdot q)(P_2\cdot q)}{P_1\cdot P_2}\right) \tag{10.12}$$

Thus, for example, the form factor for the proton (assuming no spin factors) where $P_1^{\ 2} = P_2^{\ 2} = M^2$, $(P_1 - P_1)^2 = q^2$ comes out

$$F = \frac{1}{(1 - q^2/2M^2)^2}\exp\frac{1}{\Omega}\ \frac{q^2}{1 - q^2/2M^2} \tag{10.13}$$

it is seen that for large (negative q^2) this behaves as $1/q^4$ as does the real form factor, and it generally fits rather well!

For the photoelectric matrix elements ($q^2 = 0$) assuming the states observed are those whose time state is not excited, all the matrix elements are nearly exactly the same as FKR (expect for an occasional factor of $M_1M_2/P_1\cdot P_2 = \frac{2M_1M_2}{M_1^2+M_2^2}$ which is close to 1) except they all carry the form factor F, it is theoretically much more satisfying than FKR because it is no longer chosen arbitrarily) which in this case works out to be

$$\left(\frac{2M_1M_2}{M_1^2+M_2^2}\right)^2 \exp -\frac{4}{\Omega}\frac{M_1^2Q^2}{M_1^2+M_2^2} \tag{10.14}$$

which is much like FKR empirical F except for the 4 in the exponent (so it is very close to FKR's F to the fourth power). This cuts off much too fast and makes the poor fit of the 1688 even much more unsatisfactory and also destroys other cases that fitted well.

c) Extension to higher order perturbation, in particular matrix elements of the succession of two current operators. Scattering of photon, electromagnetic mass differences or non-leptonic weak decays require the matrix element of the product of two currents. In the latter case the two currents are at the same space time point and we can prove that if the quarks are Bose the famous $\Delta I = 1/2$ rule for these decays is explained. A natural suggestion is the second order perturbation $\delta K \frac{1}{K} \delta K$. However, in summing over all states in the propagator $1/K$ what should we do?

α) Sum over all states including time states with negative norn, and include no arbitrary form factors.

β) Sum over only physical states (no time states excited) using some assumed form factor, - if so, what factor?

γ) Use the system of Fujimara et al. summing over all states including those of negative energy.

There are awkward problems associated with each way (will the commutators vanish outside the light cone?) and nothing precise or definite has yet been proposed to compare to experiment. We shall discuss this problem again when we have developed further theoretical tools, e.g. dispersion relations.

II Modofications of the Specific Model of the Theory.

Can the theory be modified to be in greater degree in agreement with experiment? The best clue to start with is probably the real value of the mass^2, which really do depend on the SU_3 and spin character. Perhaps the most important clue is the large separation of the π, η, η' system contrasted with the small differences (all accounted for by the extra mass^2 for strangeness) of ρ, ω, ϕ. It seems to indicate the interaction force is not a vector, but an axial vector. An interesting problem is the split of Σ and Λ in the baryon octets. The simple quark model predicts them to be degenerate.

Next comes the corrections that the quark model implies to itself for physical consistency. For example suppose, using PCAC or a modified quark model that has some kind of direct coupling, that a pion, Δ, and P are coupled - we can calculate an amplitude for

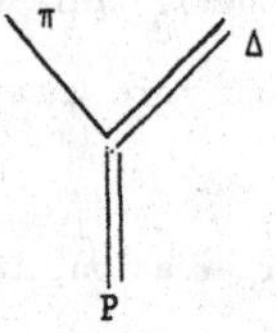

This implies, of course, that the Δ can decay into π, P and thus the Δ has a width. But in the original theory the Δ was an exact energy level. We can get the width by imagining the delta propagates also via vertical diagrams like

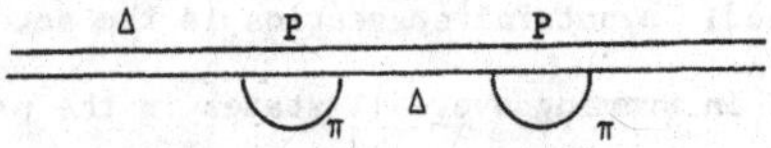

the quark model gives the coupling to be used later as the beginning of some perturbation expansion. This successfully induces an imaginary part in the propagator $i\Gamma m$ - but at the same time corrects the real part so the experimental mass is not the uncorrected mass from the direct use of the quark model. How much is the correction and can we use it to understand some part of say the Σ,Λ mass difference, the $^4(10)$ to $^2(8)$ split in the $[56, 0]$ etc? The perturbation theory is hard to use, it has a tendency to diverge, it seems to involve couplings far away from their mass shell, there are many more states than just the $P\pi$ available to the Δ, etc. Finally and most important - maybe we are partly counting the same thing twice - perhaps in the equation for the quarks we started with already contains some of the physics of interaction that we have later put into the perturbation correction.

Due to these uncertainties nobody has worked these things out. (For another example, how much is the magnetic moment of the proton corrected? It has some amplitude to be a delta and a pion, or even a proton and a pion).

This question is very serious. What is the quark model a model for? For the calculation of first order couplings among a set of ideal states which is later to be used in some definite prescription (for example, a series of diagrams like a field theory perturbation expansion) to compute the final hadron world?

The quark theory is an incomplete theory. No definite prescription is given as to what to do next to correct it, or even to use it completely in a correct and consistent (with unitarity, for example) manner. As such it reaches a dead end. It shows that certain regularities exist among the hadron states, but a precise statement of what this means or what this really implies one should include in designing a more adequate theory is still lacking. It is a most important problem in the theory of the strong interactions.

This "dead end" is a) a result of our lack of imagination of how to go further and b) a problem that every "narrow resonance approximation" has to face.

Pseudoscalar Meson Photoproduction, High Energy

Lecture 11

Pseudoscalar Meson Photoproduction - Higher Energies

We now consider what happens at higher energy - in particular from 5 to 18 GeV (as full data in the transition region 2 to 4 GeV is lacking). Here the analysis directly by sums of resonances is hopeless as too many are involved. A full analysis of amplitudes does not exist, mainly cross sections and a few asymmetry measurements with polarized photons, or polarized targets does exist but our theoretical skill at analyzing the amplitude for each helicity etc. is not yet good. Therefore this will be more of an empirical survey pointing out problems, rather than a complete theoretical analysis - in short, we do not understand this region.

References

Diebold "High Energy Photoproduction", High energy physics conference, Boulder, Colorado (1969).

Wiik "Photoproduction of pseudoscalar mesons", International symposium on electron and photon interactions at high energies, Cornell (1971).

We shall study in this part only a very small fraction of the total photon nucleon cross section at high energy. The cross section for $\gamma p \rightarrow \pi^+ n$ is, for example, 0.6% of the total at 5 GeV.

The total cross section for γ on p shows bumps due to the 1236, 1560 and 1700 resonances at low energies. It is mostly flat above 3 GeV, it falls from 125 μb to 113 μb, from 5 to 15 GeV (this is about 1/220 of πp cross section).

The reactions go increasingly in the forward direction so that it is convenient to plot them as functions of t for given s(or k, the photon momentum in the lab.). The slope of the t dependence varies relatively slowly with k, and certain characteristic regions for t are noticed. There is also a peak in the backward direction, small u, where it is best to plot u for given s.

An empirical rule, not precise, but very good for comparing data at various energies is that all the cross sections vary nearly as k^{-2} in the forward direction, and as k^{-3} in the backward direction. The reason for this is not well understood. That is, $k^2\, d\sigma/dt$ is a nearly fixed function of t. Below t about 1 $(\text{GeV})^2$ the curves for different reactions are each different. They are given on a chart from Diebold's report (reproduced here in fig. 12-1). (This is not precise data, variations of f(t) with s are smoothed over - to get exact numbers see detailed experimental reports - but they are excellent representations of things for a survey such as this.)

Above t = 1 all reactions agree in falling very much like e^{3t}. The cause of this remains unknown. In comparing to hadron collisions this rule is seen to be most accurate for photon reactions but is seen also in many strong interaction cases, for example in forward ($\pi^- p \rightarrow \pi^0 n$ or $\pi^+ p \rightarrow k^+ \Sigma^+$) as well as in backward ($\pi^+ p \rightarrow p \pi^+$) (G. Fox). It may be related to another strong interaction rule; that in high energy inclusive reactions making mesons or baryons these particles are distributed in transverse momentum $P_\perp$ as $e^{3P_\perp^2}$. This is also true of photon induced inclusive reactions. This is only true in a rough sense, $e^{-8P_\perp^2}$ for small $p_\perp^2$ to $e^{-2.2P_\perp^2}$ for large $P_\perp^2$.

The theory of transverse momentum distributions at high energy and high t is wide open. It is a very good problem to work on.

Very suprisingly backward photoproduction peaks are the only ones that seemingly do not obey the rule. They are very flat - flatter than anything else, varying as about $e^{1.2\,u}$. (This is not true for $\gamma p \to \Delta^{++}\pi^-$ which has a rapid fall off with u in the backward direction, more like e^{3u}.)

The behaviour at smaller t (t<1) can qualitatively be explained by the idea of the exchanges of Regge poles of the following trajectories:

π, and strangeness analogue K

ρ and nearly degenerate A_2

ω

K* and nearly degenerate K**

The k^{-2} behaviour would not be expected, for the various trajectories have different intercepts and slopes. But the k^{-2} cannot be a true asymptotic rule about amplitudes - it must be a sort of accidental result of several terms varying nearly that way - although quantitatively it is always a bit of trouble to get it out. This can be seen because if the amplitude varied truly as s^{-1}, or rather, in correct normalization, were constant, the photon coupling via isoscalar and isovector would differ in phase by 90° and would not interfere. Therefore we would expect $d\sigma/dt(\gamma p \to \pi^+ n) = d\sigma/dt(\gamma n \to \pi^- p)$ and they certainly are not equal.

The reason isoscalar and isovector differ in phase by 90° is this. For any amplitude f(s) the antiparticle amplitude is $f(e^{i\pi}s)^*$. Thus if $f(s) = \beta s^{\alpha(t)}$ for antiparticles it is $\beta^* e^{-i\pi\alpha} s^{\alpha}$. But for an isoscalar coupling the amplitudes for particle and antiparticle are opposite, this requires $\beta_s = ic \exp(-i\pi\alpha/2)$, c is real. For isovector coupling the amplitudes are the same because the antiparticle has opposite I-spin thus $\beta_v = c' \exp(-i\pi\alpha/2)$, c' is real. This shows that β_s and β_v differ in phase by 90°.

In Regge theory the exchange of a trajectory can be interpreted as a sum over t channel resonances. In each case the polarization of the exchanged resonance is used up in getting as high a power of s as possible, by being in the direction of the two longitudinal momenta. Take, for example, the case where a vector particle is exchanged and there is no spin flip at either vertex (t fixed high s $P_1 \approx P_3$, $P_2 \approx P_4$)

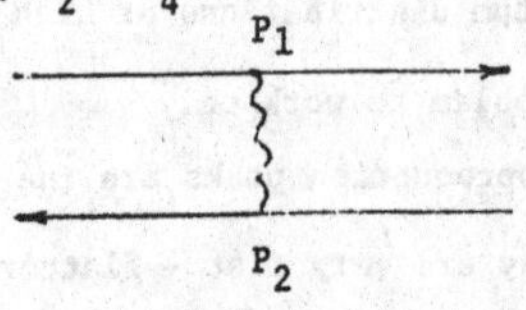

The amplitude goes like $\sum_{Pol}(P_1\cdot e)(P_2\cdot e) = P_1\cdot P_2 = s$. The spin is therefore not available to carry longitudinal angular momentum. If there is a helicity flip $\Delta\lambda$ at one of the vertices, the spin of the exchange is again used up in the term with the highest power of s; hence there is a factor $Q^{\Delta\lambda}$ or $\sqrt{t}^{|\Delta\lambda|}$. If both vertices have a flip the amplitude goes as $\sqrt{t}^{|\Delta\lambda_1| + |\Delta\lambda_2|}$ ideally. This later is true if the $\Delta\lambda$ are of opposite sign, it is required by the overall conservation of angular momentum. However if the $\Delta\lambda$ are equal, say to 1, we expect t but this is not required by angular momentum conservation. In fact, in such cases the effect is lost by absorption. Absorption corrections for small impact parameters reduce the t to its lowest physical possibility (in this example to a constant). Only if the couplings really factorize must it be $t^{|\Delta\lambda_1| + |\Delta\lambda_2|}$ otherwise absorption can reduce the power.

Lecture 12

Pseudoscalar Meson Photoproduction - Higher Energies (continued)

We now discuss each case in closer detail (following G. Fox) $\gamma p \to \pi^- \Delta^{++}$

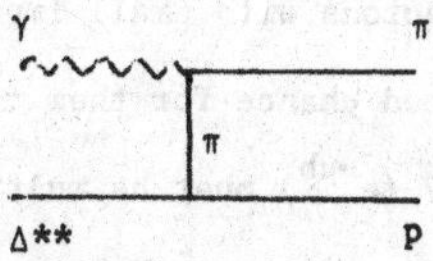

This case can be explained simply by π exchange. At the photon to π vertex we must have a flip of helicity, at the $p\Delta$ vertex none is necessary (because of the p-Δ mass difference); hence the amplitude is $\sqrt{t}/(t-m_\pi^2)$ showing a dip at $t = 0$, a maximum at $t = m_\pi^2$. π exchange expects k^{-2} cross sections ($s^{2\alpha-2}$, $\alpha \approx 0$). Gauge invariant one pion exchange with absorption corrections works fine in this case.

$\gamma p \to \pi^+ n$, $\gamma n \to \pi^- p$

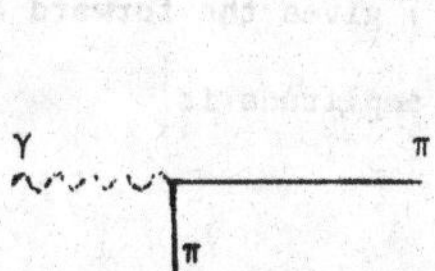

We have a helicity flip at the $\gamma\pi$ vertex and one (from parity) at the nucleon vertex. Hence ideally we expect $t/(t-m_\pi^2)$, there should be a drop at $t = 0$ but if we look at the data in figure 12-1 we see there is a spike in the forward direction. The answer to this is: "because of absorption". To understand how absorption affects the amplitude, perform a Fourier transform to the impact parameter representation. The amplitude to find a pion now looks like $\nabla\ (e^{-\mu b})$ which is an odd function of b. The photon couples to this pion with amplitude that goes like $\nabla\nabla\ (e^{-\mu b})$, the first gradient is because the pion is in a P wave with respect to the nucleon, the second gradient is because the photon couples to the pion in a P wave. $\nabla\nabla\ (e^{-\mu b})$ is a function that looks like

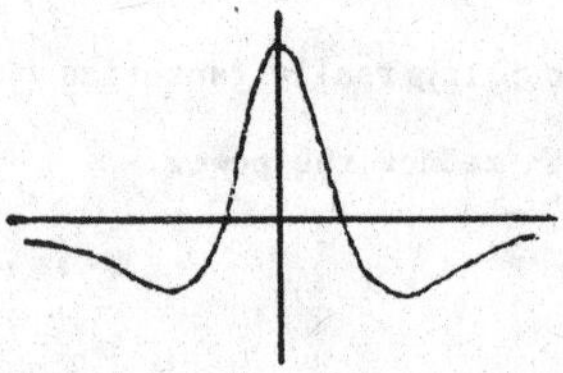

The forward amplitude is proportional to the integral of this function which is zero so we still haven't explained the observed peak. Absorption is here introduced, pions which are scattered by photons with small impact parameter do not always get out because there is a good chance for them to interact strongly with the nucleon. The function $\nabla\nabla\ (e^{-\mu b})$ must be multiplied by a function a(b) like:

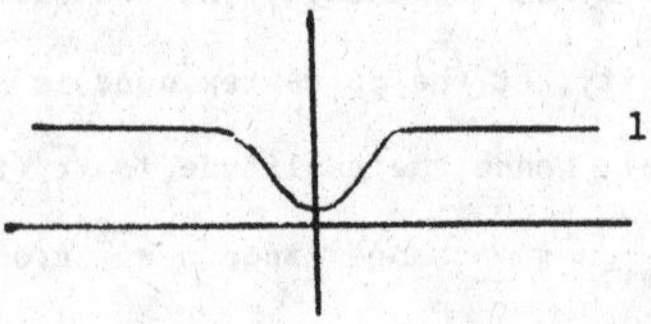

The integral of the product $a(b)\ \nabla\nabla\ (e^{-\mu b})$ gives the forward amplitude which is now non zero. The result is that the amplitude is

$$\frac{t}{t-m_\pi^2} - C \qquad (12.1)$$

G. Fox has noticed the empirical rule that setting C = 1 works very well in many cases. The amplitude becomes $m_\pi^2/(t-m_\pi^2)$. (Such peaks are common in strong interactions, e.g., p n change exchange pn→np).

The difference between $d\sigma/dt(\gamma p \to \pi^+ n)$ and $d\sigma/dt(\gamma n \to \pi^- p)$ shows the presence of ρ-like exchange. The amplitudes in the two cases go as

$$\gamma p \to \pi^+ n = \pi + \rho + A_2$$

$$\gamma n \to \pi^- p = \pi - \rho + A_2$$

The π and ρ interfere coherently because the ρ coupling is mainly due to spin flip (see below). The sign of coupling agrees also with what is expected by naive quark model. (It also agrees with the sign in a similar strong interaction case $np \to pn = \pi + \rho$ and $p\bar{p} \to n\bar{n} = \pi - \rho$.) The expected energy dependence for various t for the ρ is not clearly seen. Perhaps for small t pions dominate. For t near −0.5 a dip is expected in $\gamma p \to \pi^0 p$ by the fact that $\alpha(\rho) = 0$ there (roughly $\alpha(\rho) = 0.5+t$). In Regge theory the amplitude for a ρ trajectory exchange is proportional to

$$\frac{(1-e^{-i\pi\alpha})}{\sin\pi\alpha\ \Gamma(\alpha)}\, s^{\alpha(t)-1}\ . \tag{12.2}$$

The pole in $(\sin\pi\alpha)^{-1}$ at $\alpha=0$ is "killed twice" by the phase in the numerator and by $(\Gamma(\alpha))^{-1}$, so the contribution of the ρ trajectory vanishes at t = −0.5. The dip is seen more clearly in $\pi^+ p \to \pi^o n$; in $\gamma p \to \pi^o p$ the zero being in a double flip amplitude may be obscured by absorption. In a single flip amplitude it should show up.

Asymmetry for $\gamma p \to \pi^+ n$

$$A = \frac{\sigma_\perp - \sigma_\parallel}{\sigma_\perp + \sigma_\parallel} \tag{12.3}$$

$\sigma_\perp$ and $\sigma_\parallel$ are the cross section for photons polarized perpendicular and parallel to the plane of scattering respectively. Experimentally A looks like

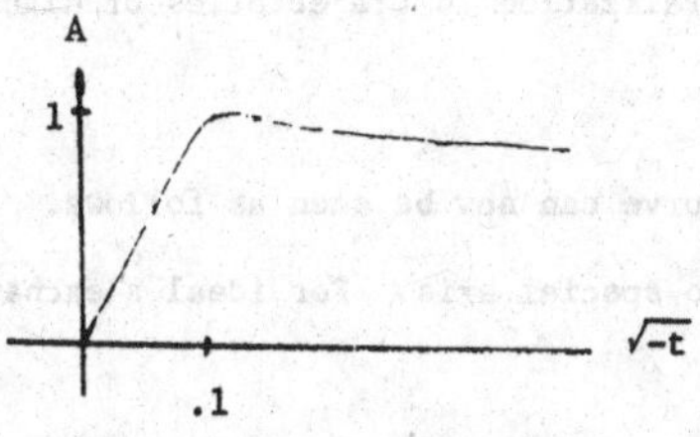

for small t $\sigma_{\parallel} \approx \sigma_{\perp}$, for $\sqrt{-t} > .2$ $\sigma_{\parallel}$ is small. This can be explained by considering the exchanges which contribute to $\sigma_{\perp}$ and $\sigma_{\parallel}$. $\sigma_{\perp}$ is contributed only by natural parity exchange ($P = (-1)^J$ e.g. ρ,ω) $\sigma_{\parallel}$ is contributed only be unnatural parity exchange ($P = -(-1)^J$ e.g. π).

To see this consider the following picture:

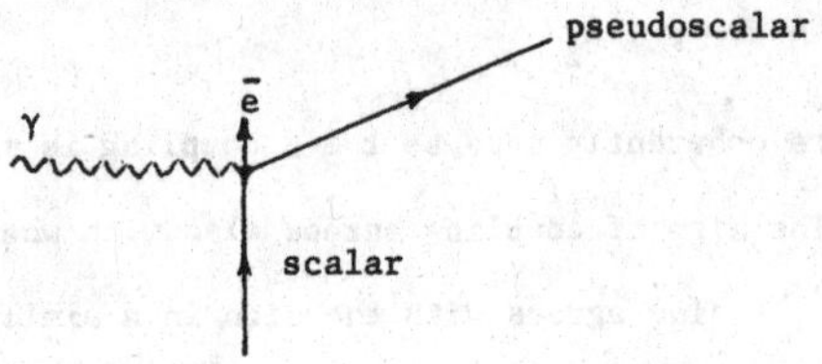

The photon comes in with polarization $\bar{e}$ in plane of sheet, if this is reflected in a mirror in the plane of scattering everything looks the same except that since there is a pseudoscalar particle the amplitude changes sign, it is therefore zero.

Consider now a photon coming in with perpendicular polarization:

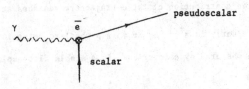

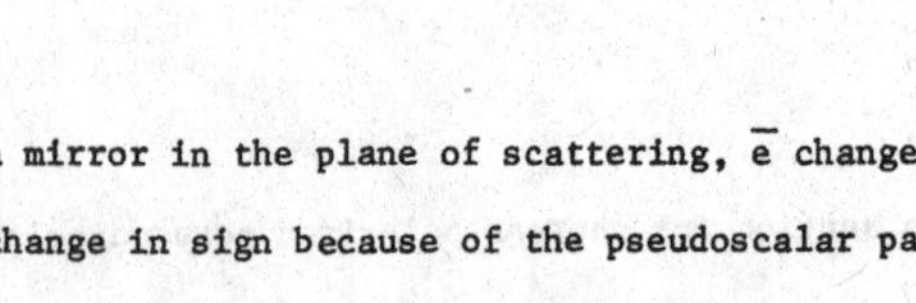

Again reflect in a mirror in the plane of scattering, $\bar{e}$ changes sign, but there is another change in sign because of the pseudoscalar particle so a natural parity exchange can contribute to $\sigma_{\perp}$. Similar arguments can be used in the case of pion exchange, generalization to trajectories of higher angular momentum gives our rule.

The reason for the asymmetry curve can now be seen as follows. At $t = 0$ $\sigma_{\parallel} = \sigma_{\perp}$ by geometry, as there is no special axis. For ideal π exchange we have the amplitudes

$$A_{\parallel} = \frac{t}{t-m_\pi^2} \qquad (12.4)$$

$$A_{\perp} = 0$$

for ρ we have $A_{\parallel} = 0$ $A_{\perp}^{\rho} \neq 0$. From absorption (t=0 correction) we must have $A_{\parallel} = -1/2\ A_{\perp} = -1/2$ to get zero cross section using G. Fox's rule for size. The total is

$$A_{\parallel} = \frac{t}{t-m_\pi^2} - \frac{1}{2} = \frac{1}{2}\left(\frac{t+m_\pi^2}{t-m_\pi^2}\right) \qquad (12.5)$$

$$A_{\perp} = -\frac{1}{2} + A_{\perp}^{\rho}$$

$A_{\perp}^{\rho}$ should go to zero at small t. Neglecting it we expect the asymmetry

$$A = \frac{|A_{\perp}|^2 - |A_{\parallel}|^2}{|A_{\perp}|^2 + |A_{\parallel}|^2} \qquad (12.6)$$

to start at zero at $t = 0$, rise rapidly to 1 at $t = -m_\pi^2$ (which it does experimentally) and to fall off toward zero for higher t. It does fall off slightly, to perhaps 0.8, but not as much as expected; presumably other terms like $A_{\perp}^{\rho}$ are beginning to contribute strongly.

$\gamma p \to \pi^0 p$

$d\sigma/dt$ for this reaction shows a very sharp forward peak, it is due to photon exchange (called Primakoff effect), the diagram is

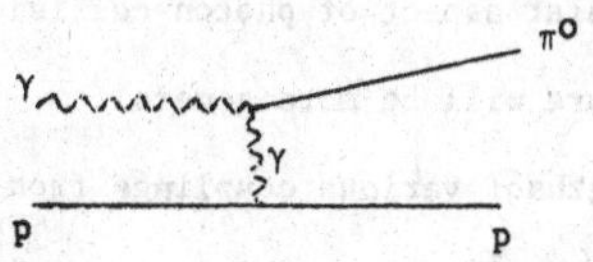

we know that π^0 couples to two photons from its decay. In fact, this effect provides the most accurate way to measure the lifetime for $\pi^0 \to \gamma\gamma$; the exchanged photon in the Primakoff effect is almost on its mass shell.

For t outside the forward peak, the reaction can come from some exchange - we expect the ω to predominate (see below) and to have a strong single flip (flip at $\gamma\pi$ vertex, non-flip at nucleon vertex). Thus it vanishes at $t = 0$, and does show the dip when $\alpha(t) = 0$ at $t = 0.5$. The energy dependence is entirely wrong for small t. A plot of experimental $\alpha(t)$ (from $d\sigma/dt = s^{2\alpha(t)-2}$) shows α near 0 ± .2 for all t out to -1.2 where it may be -1. If ω we expect $\gamma n \to \pi^o n$ to be equal. The ratio $\frac{\gamma n \to \pi^o n}{\gamma p \to \pi^o p}$ is 1 for small t, falls to 0.6 near t = 0.5 and rises again to 1 at large t.

The asymmetry for polarized photon parallel $\sigma_{\parallel}$ and perpendicular $\sigma_{\perp}$ to the production plane has also been measured. $A = (\sigma_{\perp} - \sigma_{\parallel})/(\sigma_{\perp} + \sigma_{\parallel})$. The cross section is predominantly $\sigma_{\perp}$ except near the dip when $\sigma_{\parallel}$ may be as much as 20%.

$\gamma p \to K + \Lambda$ $\gamma p \to K + \Sigma$

These reactions are due to K* and K** exchange. A bit of K exchange at lower t makes Λ greater than the Σ because the coupling $KP\Lambda$ is bigger than $KP\Sigma$ (see below).

$\gamma p \to \eta p$

This is dominated by ρ exchange instead of ω exchange, so no dip is seen at -0.5 because of the double flip.

Quantitative fits have been attempted along these lines, but each needs several complicating features like absorption, Regge cuts etc. It is not worth our while to follow any of them in specific detail for no one model automatically easily gives detailed fits to all the reactions.

What we can say however, is that in this large s region (at least for small t) the concept of t channel exchanges seems to be a guiding principle that is fairly successful. This same effect is seen in hadron collisions in even more detail; it is not a particular aspect of photon collisions. Therefore the discussion of the next lecture will be more general.

We derive now the relative strengthsof various couplings from the quark model. The arguments above use these results.

To get these relative strengths we assume mesons couple through currents. The currents are obtained by taking the meson quark wavefunctions and interpreting the quark states as creation and annihilation operators.

We have $\rho = u\bar{u} - d\bar{d}$, $\omega = u\bar{u} + d\bar{d}$; hence ρ couples to a nucleon like a current with u charge +1, d charge -1. ω couples like u charge +1, d charge +1. A photon couples like u charge 2/3, d charge -1/3 or like $\frac{1}{6}\omega + \frac{1}{2}\rho$. To a neutron the coupling of the ρ is reversed in sign, the ω is not. Therefore $\gamma p = \frac{1}{2}\rho + \frac{1}{6}\omega$ $\gamma n = -\frac{1}{2}\rho + \frac{1}{6}\omega$ which gives $\rho = \gamma p - \gamma n$, $\omega = 3(\gamma n + \gamma p)$.

The couplings go like

	to charge	to magnetic moment
γp	1	3
γn	0	-2
ρ	1	5
ω	3	3

Therefore ρ couples mainly to spin, the flip amplitude predominates but for the ω no such predominance occurs.

From $\pi^o = u\bar{u} - d\bar{d}$ we get the couplings of ω, ρ to γ and π^o. Hence between π^o and ρ^o (which have the same $u\bar{u}$ and $d\bar{d}$ signs) coupling the photon to u plus photon to $d = \frac{2}{3} + (-\frac{1}{3}) = \frac{1}{3}$, to ω it is coupling to u minus coupling to $d = \frac{2}{3} -(-\frac{1}{3}) = 1$, therefore

$$\frac{\pi^o \omega\gamma}{\pi^o \rho\gamma} = \frac{3}{1}$$

For the $\eta = \frac{1}{\sqrt{6}}$ $(2 s\bar{s} - u\bar{u} - d\bar{d})$ the ratio is reversed

$$\frac{\eta\omega\gamma}{\eta\rho\gamma} = \frac{1}{3}$$

Pseudoscalar mesons couple to nucleon octet via SU_3 with F = 2, D = 3, thus

$$\frac{K\Sigma N}{K\Lambda N} = \frac{\frac{1}{\sqrt{2}}(F-D)}{\sqrt{\frac{3}{2}}\,(F+\frac{1}{3}D)} = \frac{-1}{3\sqrt{3}} \ .$$

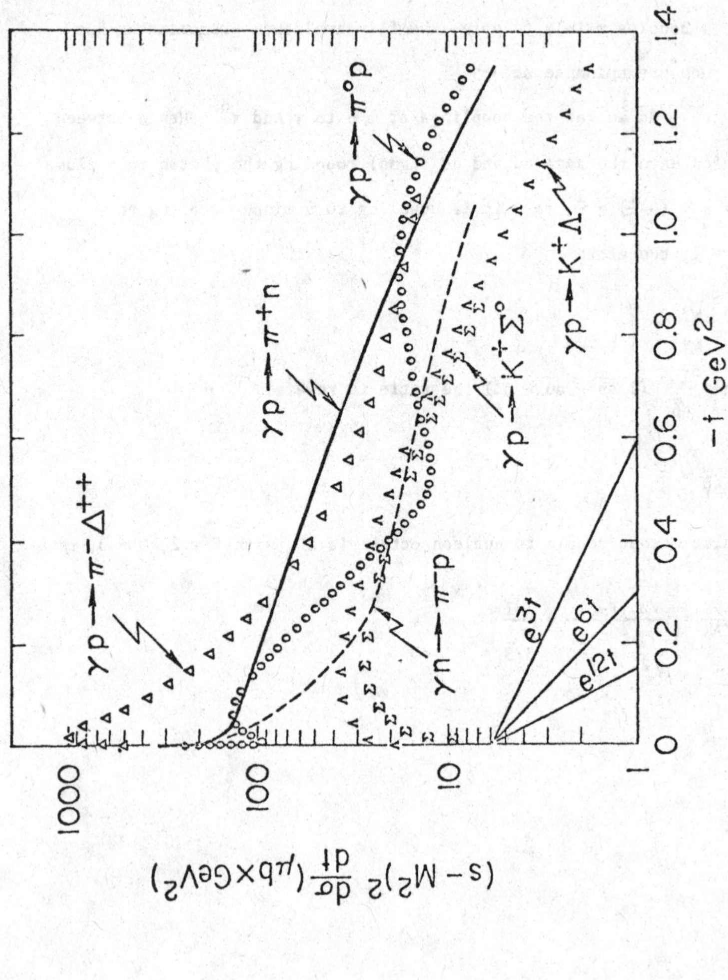

$(s - M^2)^2 \frac{d\sigma}{dt} (\mu b \times GeV^2)$
1000
100
10
1
0
0.2
0.4
0.6
0.8
1.0
1.2
1.4
$-t$ GeV^2
$\gamma p \to \pi^- \Delta^{++}$
$\gamma p \to \pi^+ n$
$\gamma p \to \pi^0 p$
$\gamma n \to \pi^- p$
$\gamma p \to K^+ \Sigma^0$
$\gamma p \to K^+ \Lambda$
e^{3t}
e^{6t}
e^{12t}

FIGURE 12.1

t-Channel Exchange Phenomena

Lecture 13

<u>t-Channel Exchange Phenomena</u>

There have been a number of theoretical views as to how to understand these t channel exchange effects and I will discuss a number of them. They are closely interrelated and are undoubtedly to a large extent, different ways of describing the same thing. I will give a very brief discussion of each, indicating their relationships to each other and their various degrees of incompleteness. Some of the ways of understanding t exchanges at large s are:

a) The geometrical or impact parameter viewpoint

b) The Regge pole formula

c) t-channel resonances

d) s-channel resonances.

The first view (has been discussed most completely by Harari) notes that as $s\to\infty$, the angular momenta can get very high so that if the ℓth orbital momentum is selected where $\ell = kb$ is kept fixed as s increases we can define b ever more accurately. This b ultimately becomes the 2-dimensional space vector perpendicular to the plane of collision and is the Fourier transform variable corresponding to $Q_\perp$ the transverse momentum transfer vector.

Amplitudes $A(Q_\perp, s)$ can therefore be written as

$$A(Q_\perp, s) = \int e^{i\bar{Q}_\perp \cdot \bar{b}}\, a(\bar{b}, s)\, d^2\bar{b} \qquad (13.1)$$

and the behaviour of A is studied by describing $a(\bar{b}, s)$ using geometrical intuition to assist in understanding $a(\bar{b}, s)$ rather than trying to gather intuition of $A(Q_\perp, s)$ directly.

Diffraction is readily understood by supposing there is an amplitude for a particle to pass through a target without interacting which is very small for small b, but rises to one outside the target.

$A(\bar{b}, s)$ would look like

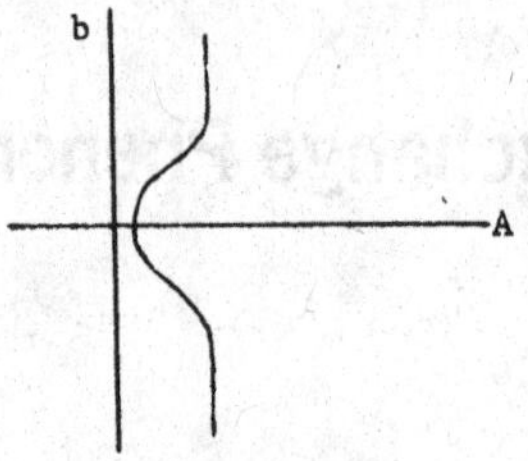

The energy dependence of this amplitude (nearly independent of energy) is taken from experiment. (It is vaguely understood as a strong interaction of an object of fixed size.)

A reaction like charge exchange, say $\pi^- p \to \pi^0 n$ is described by supposing, at fixed b that the π^- going through has some amplitude to change to a π^0. (The energy dependence of this amplitude may again be guessed from experiment, or from another theory like Regge pole theory.) This amplitude is largest in a ring

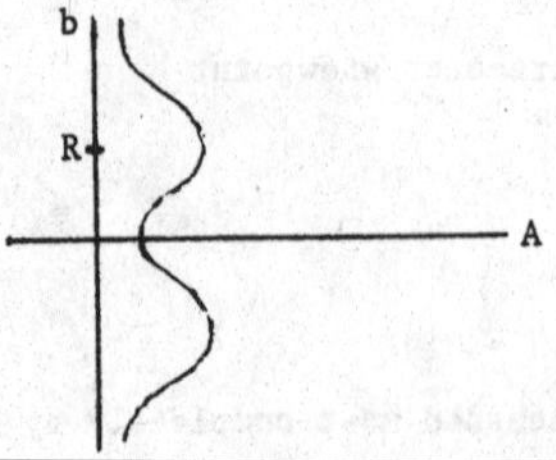

because for large b no interaction occurs while for small b there is little chance of getting through. In crudest approximation we can say all comes from a ring at radius R, leading in Fourier transform to an amplitude like $J_0(\sqrt{-t}\, R)$, for no helicity flip. This expects a dip at t = -0.15 which is sometimes found.

For a single helicity flip where the amplitude must vary across the impact parameter plane as $e^{i\phi}$, the result is $J_1(\sqrt{-t}R)$. For R = 1 fermi this has a zero at t = -0.5 and is, in this model the origin of the dip at this t in pion exchange scattering, photon production of π^0 etc. (The dip that in Regge theory we associated with the zero of $\alpha(t)$).

This model does give, with some exceptions, a rough description of the general behaviour, but it is not a complete theory because the energy dependence of amplitudes is not understood, nor are the coupling strengths.

The view that a particle is a point traveling at a fixed impact parameter rather than a structure with finite dimension in the transverse plane (i.e. two pie plates colliding) is an oversimplification. Attempts to deal with this have so far, lead to complications and ambiguities and no clear definite picture has emerged.

b) The Regge pole formulas result from a suggestion (motivated by an analysis of non-relativistic scattering by Regge) that the amplitude function for two body reaction A(s,t) be studied as an analytic function of the energy s and an angular momentum variable in the t channel, α. For large s the analytic formula was suggested

$$A(s,t) = (\sum_i + \int_i) \beta_i(t) s^{\alpha_i(t)} \qquad (13.2)$$

where a sum over special values of i (called poles) and a continuum of i values (called cuts) is in general expected. Nearly any function can be so expanded but the usefulness of the idea depends on making further assumptions. Evidently as $s\to\infty$ the largest α's would determine A, and it was hoped that the poles would have these, definitely above the α values where the cuts start, so A would behave as $\beta(t)\, s^{\alpha(t)}$ or a sum of a very few terms of this form. If so, $\beta(t)$ would be factorizable $\beta_{AC}(t)\, \beta_{BD}(t)$ in a reaction A + B → C + D, and $\alpha(t)$ if extended to positive t would be an integer for those t equal to the mass of a real particle of appropriate angular momentum. It is only at such a t that A could have a pole like singularity in t.

For certain reactions (e.g. $\pi^- p\to\pi^0 n$) these expectations have been brilliantly confirmed, but generally complications set in. For example, for pion exchange we saw how absorption, a concept from viewpoint (a) often modifies the simple pole picture. In the "pure" Regge view any modification of the simple pole formula for A(s,t) can be described, of course, by adding cuts - but what cuts to add, and how strongly requires outside ideas, like borrowings from the absorption model. The cuts start, for large s, right at the position of the poles and cannot be easily separated from them.

It is found empirically that $\alpha(t)$ is nearly a straight line. The reason for this is unknown.

Also to describe the nearly constant total cross sections a special trajectory was invented, the pomeron, with a special value one for α at $t = 0$, without any dynamic explanation of this part.

This theory is again incomplete as no information is supplied for getting $\beta(t)$, and, in practice more important, for determining the rise and behavior of cuts. Attempts to avoid using the absorption model ("conspiracies," etc.) have failed.

Obviously some kind of integration of the ideas of a, b will be the most profitable.

c) The relation of the Regge form to the particles on the trajectory can be looked at another way (van Hove). Suppose we have a theory which would give a sequence of particles of ever increasing angular momentum. Then the Born terms representing first order exchange of all these t- channel particles will give a Regge-like sum. I give an example. Let the mass^2 go as $m_n^2 = m_o^2 + n\Omega$, where m_o^2 is the mass^2 of a scalar, $m_o^2 + \Omega$ of a vector, $m_o^2 + 2\Omega$ of a tensor (spin 2) particle, etc. The exchange of these for a reaction $A + B \rightarrow C + D$ where for simplicity A, B, C, D are scalar, goes as

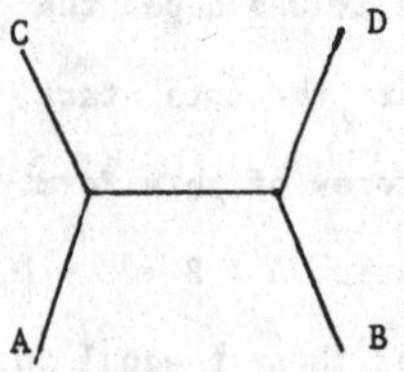

$$C_o \frac{1}{t-m_o^2} - C_1 \frac{s}{t-m_1^2} + C_2 \frac{s^2}{t-m_3^2} + \ldots \qquad (13.3)$$

The s for vector comes from $(P_A + P_C) \cdot e$ at the upper coupling and $(P_B + P_D) \cdot e$ at the lower summed over e. Likewise s^2 for the tensor comes if we use the full two tensor indices to carry information about $P_B + P_D$ from vertex to couple to $P_A + P_C$ to make $((P_A + P_C)\cdot(P_B + P_D))^2$ or s^2 in the limit. C_n are squares of coupling constants. Let $m_n^2 = m_o^2 + n\Omega$, and let C_n equal $1/n!$ say, our sum is

$$A = \sum_{n=0}^{\infty} \frac{1}{n!} \frac{(-s)^n}{(t-m_o^2-n\Omega)} = -\sum_{n=0}^{\infty} \int_0^{\infty} \frac{(-s)^n}{n!} \exp\,(t-m_o^2-n\Omega)x\,dx$$

$$= -\int_0^{\infty} e^{(t-m_o^2)\,x} \exp\,(-se^{-\Omega x})dx$$

setting $u = se^{-\Omega x}$ and $(t-m_o^2)/\Omega = \alpha(t)$ the integral can be transformed to

$$A = -\frac{1}{\Omega}\; s^{\alpha(t)} \int_0^s u^{-\alpha(t)-1}\, e^{-u}\, du \; .$$

For large s the upper limit on the integral can be put ∞, to get the asymptotic form

$$-\frac{1}{\Omega}\, s^{\alpha(t)}\; \Gamma(-\alpha\ (t)) \qquad (13.4)$$

The poles and residues of this function $\Gamma(-n+\varepsilon) = \frac{(-1)^n}{n!\,\varepsilon}$ reproduce the original series. In general any Regge term can be expanded by its poles in t (coming from $\beta(t)$ and the series is then a van Hove series determining particles and couplings. Or some guess can be made of states and residues and a Regge term derived. These two forms are mathematical expressions of the same theory, and permit different points of view of the same thing; sometimes one, sometimes another is more suggestive of a new idea.

<u>Comments</u>

For negative s the series (13.3) gives an amplitude rising as e^s. It is suggested that the coefficients cannot fall as quickly as $1/n!$ Alternatively one can say that the values for negative s are taken from the asymptotic form for positive s by analytic continuation.

Clearly in this sum, for negative t, we are far away from the true resonances and are again in the position of not knowing what the real sum should be unless we have a complete theory. Such analytic extension formulas are hard to use practically - for example what does observation of the first two resonances of the serial tell us about the asymptotic form?

$$\frac{1}{t-m_o^2} - \frac{s}{t-m_1^2} + ? = ?$$

If there were no other terms the asymptotic s behavior would go as s. Yet from m_0^2, m_1^2 we can get a little information on the behavior of $\alpha(t)$. How does it work? I don't know - the analytic extension of approximate functions, or empirically partially determined functions is a mathematical problem I do not understand nor trust, nor can I guess errors. To show the type of difficulty, suppose a sum of t channel resonances gives $\beta s^{\alpha(t)}$; now change the coefficient of the $\ell = 1$ term by ΔC. Surely the sum is now $\beta s^{\alpha(t)} + \frac{(\Delta C)s}{t-m_1^2}$ which for large s goes as s. The point is a modification of the second term only is not "natural" or "smooth" and changes the result drastically. Whatever physically causes the change in the first term changes all the other terms a bit in such a way that the asymptotic form is only mildly altered.

To summarize the van Hove sum is not really useful except theoretically. If you have some theory of all the t-channel resonances and their couplings then you can be assured the theory will lead to Regge asymptotic behaviour - or if you have some theory leading to Regge asymptotic behaviour you can, by looking at the poles (in t) in its expression deduce where t-channel resonances are implied ($\alpha(t)$ = integer) and something about their couplings.

Summing the t-channel resonances suggested by the harmonic oscillator quark model for meson resonances ($Q\overline{Q}$) does give an expected Regge behaviour in this manner. Detailed numerical comparisons have not yet been made.

Lecture 14

d) s-Channel Resonances

Another theoretical idea is that all of the behaviour at high energy also can be understood by just continuing our s channel resonance analysis (e.g. by Walker) to higher energy. A behaviour like a dip fixed in t, as s increases means a feature that appears at gradually decreasing angles $\theta = t/s$ so that a sharper and sharper angular distribution results. This means that higher and higher angular momentum s-channel resonances must be contributing as s increases. But this is, of course, all right for we know higher angular momentum resonances are available at higher s. The approximate stability of f(t) curves as s varies does imply a great deal about how the resonances and their couplings must behave at large s.

That the entire curve comes from summing s channel (and, some would say, u channel) resonances only is a tremendous restriction which, if true, might be so strong as to guide theory to the one and only exact expression that can do this - thus, much attention has been paid to this idea. Most statements are, at first, not that ambitious, because comparison to experiment again is meaningless if we have to decide on the hypothetical value of a sum of resonances far from their resonant energies. Instead it is noted that the imaginary part of a resonance $\Gamma((E-E_r)^2 + \Gamma^2/4)^{-1}$ is only large near the resonance so the principle is studied in the form

$$\text{Im}\,A(s,t) = \Sigma\ \text{Im (s-channel resonances)} \qquad (14.1)$$

except perhaps for the pomeron.

Such questions as the contribution of u channel resonances (which are always off resonance so the imaginary part is very small, zero for infinitely narrow resonances) and whether the pion exchange term (which is real) should be included are sidestepped in this way.

It must be emphasized that s-channel poles are explicitly meant. For if an integral over a continuum is also included nearly all functions A(s,t) can be so represented.

If we add the principle (duality) that the t-channel behavior (resulting from the s-channel poles) must if analytically extended, have poles at the same mass^2 for the same quantum numbers as the s-channel resonances we have such a severe restriction that perhaps only a nearly unique solution exists, and hence a solution of the strong interaction physics could result. Much work has been done in this direction - success has not been attained. Naturally the principle might be wrong -- perhaps s-channel and t-channel resonances must be separately added - or what amounts to the same thing s channel cuts are necessary if the amplitude is to correctly represent the high energy behavior in s and t. There is no physical principle underlying the duality hypotheses - it is simply a guess at a possible mathematical property of an amplitude. It is an observation at low energy that amplitudes are dominated by resonances, so that some sort of infinitely narrow resonance approximation may be a good starting point. But is this also valid at high energy where in most collisions many particles are produced? Can they be understood as the products of cascade disintegration of resonances or is there an important contribution of direct coupling to the continuum?

We remember that at high s the two body final state we have studied seems to be only a very small part of the cross section. The multibody states which constitute the major part of the collision cross section will be discussed later.

Veneziano formula

Such ideas are encouraged by the existence of a function (Veneziano) which can be expanded entirely in terms of s-channel poles - or t-channel poles, and which has Regge behavior at high energies.

$$A(s,t) = \frac{\Gamma(-\alpha(s))\Gamma(-\alpha(t))}{\Gamma(-\alpha(s)-\alpha(t))} \tag{14.2}$$

where $\alpha(s)$ is a straight line in s.

The residues at the poles in some generalizations are not necessarily always positive in cases where they should be.

This is only a possible form for A(s,t) if the scattering particles are spinless - much work without clear conclusion has been done in trying to generalize this. Also to generalize from 4-point to n-point functions.

A fundamental problem is this: This is a narrow resonance approximation. How precisely are we to correct it? (The same question we discussed for the quark model.) Again perhaps the same question in another language - this is presumably not directly the formula for A(s,t) but for a deeper amplitude that must still be, as we might say, "corrected for absorption." Do we have the Born term of some expansion?

Attempts to extend duality or Veneziano ideas to photoelectric couplings have been singularly unsuccessful. However, the problem of the mathematical behavior of photon couplings should not be disregarded. They are especially important because they contain beside s, t a new variable q^2 the mass^2 of the photon (which may for example be virtual via electron scattering). The amplitude $A(s,t,q^2)$ has analytic properties in all 3 variables s, t and also q^2 (we shall study them later) and this increase in restrictions may be very helpful in suggesting forms or theories.

Finally we can ask whether a dynamical model (such as the 3 quark model) of hadrons could be expected to lead to just the sum of resonances and couplings needed to generate mainly forward scattering. I think it is very likely to have no difficulty. We may draw the following diagram:

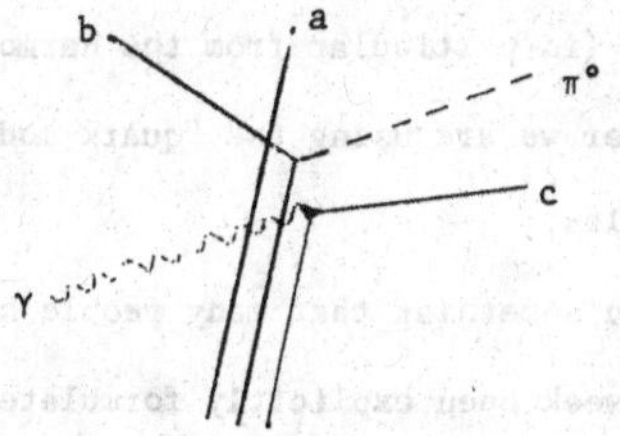

If we suppose the photon and pion each couple to one quark - but each to a different one, then quarks a and b have large momenta and the amplitude that the final configuration a, b, c is in a simple proton state, where they presumably have small relative momentum is very small. Thus at high energy the most important case is that a single quark a receives the photon and emits the pion:

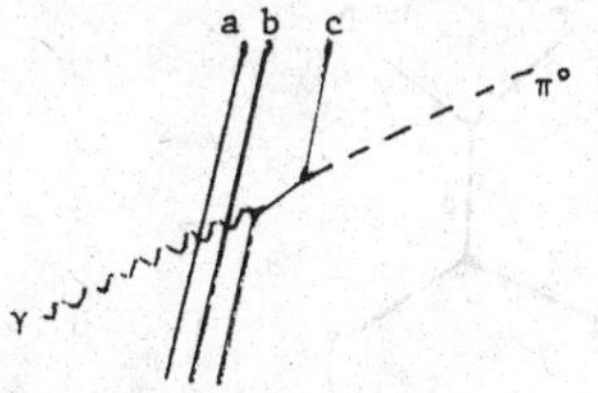

The net momentum c gets is only the momentum transfer k-q and the amplitude a b c is a simple proton is large if, and only if, this transfer is small. Therefore the pion goes forward in a way characterized by a function of the momentum transfer.

At any rate a dynamical theory of s channel resonances alone might not find it too difficult to obtain apparent t-channel exchange effects.

Estimates of Coupling Constants

In a number of cases we estimated sizes of coupling constants. For example, we said ρ-exchange was mainly flip from a nucleon. Such things come from theory outside the impact parameter, or Regge theory which left such constants undetermined. They could come from other experiments (e.g. flip is large can be got from π nucleon scattering) but nevertheless theories are available for what to expect. We used mainly SU_3, but even here F/D ratios must be known - and these are given well by SU_6, or what I have called a quark model estimate.

The "quark model" used here is really not the same as the harmonic quark model used for the s resonances applied to t resonances. We are not dealing with t resonances but rather with t trajectories. The quantitative theory of trajectories to be expected from any (in particular from the harmonic) quark model has not been worked out. Rather we are using the "quark model" just to count - to make up SU_3 and F/D rules.

Even then we implicitly supposed something that many people have used (but to our knowledge has only last week been explicitly formulated by Kislinger). We assumed the ρ (or ω) trajectory was coupled in the same proportion as currents are coupled (of the appropriate quantum numbers I = 1, 0). This idea usually comes in two steps. (1) the vector mesons on their mass shell couple like currents and (2) the ρ,ω,ϕ trajectories couple like the ρ,ω,ϕ mesons do. Stated more precisely consider the amplitudes for the following diagrams.

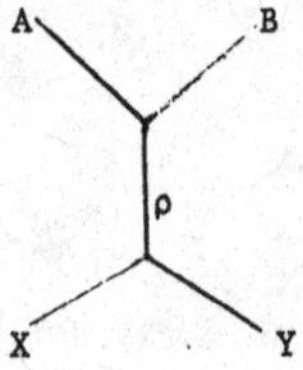

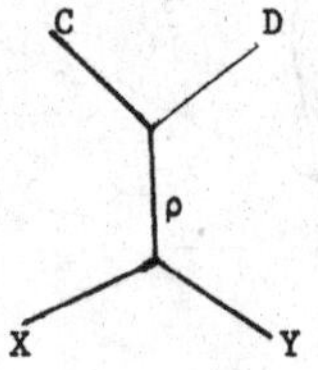

The ratio of these amplitudes is $\langle B|j|A\rangle \,/\, \langle D|j|C\rangle$ where j is the current with ρ quantum numbers. We have assumed the amplitudes have not been corrected for absorption.

Another formulation of this "quark counting" game is to draw diagrams for t exchanges between hadrons like

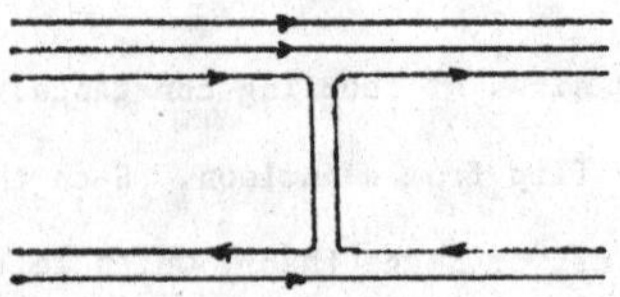

implying that baryons are QQQ and mesons are $Q\overline{Q}$ in the usual way but that the interaction can be largely understood as a sum of terms each representing the scattering of one quark in one hadron with one in another. (Lipkin).

It is clear that our interpretations of higher energy scattering combines several ideas. What is evidently lacking today is some overall view that combines these nicely together. In the past too much attention has been paid to forcing the amplitude into one mold or another rather than combining the more well established and central ideas of each of them into some synthesis.

Vector Mesons and Vector Meson Dominance Hypothesis

<u>Properties of Vector Mesons</u>

Before we discuss production of vector mesons by photons we first discuss the simpler problem of photon-vector meson coupling.

<u>Electron production of vector mesons</u>

In a colliding beam experiment we can measure $e^{+}+e^{-} \rightarrow$ hadrons. This is interpreted of course as a virtual photon of positive $q^{2} = (2E_{cm})^{2}$ and gives us a direct measure of

$$\langle \text{hadrons} \, | J_{\mu}(q) | \, 0 \rangle$$

for positive q^{2}. This matrix element is dominated by three resonances at suitable values of q^{2} corresponding to the ρ, ω, ϕ neutral mesons of spin 1^{-}, just like the photon.

For example if the hadrons is a pair of K's, $K^{+}K^{-}$ the cross section is very small unless q^{2} is close to $(1020)^{2}$ (ie m_{ϕ}^{2}) and then rises way above background in a beautiful Breit-Wigner curve of width 4.7 ± .7 MeV. The interpretation, of course, is that we have gone through a resonance, the ϕ via

a diagram like

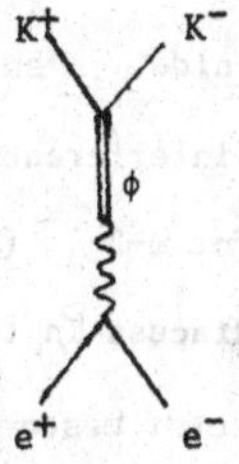

To analyze this we have two amplitudes, one purely hadronic $\langle K^+K^-|\phi\rangle$ and a photon amplitude which we write as $\sqrt{4\pi e^2}\; e_\mu \langle\phi|J_\mu|0\rangle$ at the mass^2 of the ϕ, e_μ is the photon polarization. The second factor may be written $\langle\phi|J_\mu|0\rangle = F_\phi e_\mu$ where e_μ is the polarization vector of the vector particle.

(There is no generally accepted convention for how to write F_ϕ. One way is to write $m_\phi^2/g_\phi = F_\phi$ for what we have called F_ϕ, others write $m_\phi^2/2\gamma_\phi$ so γ_ϕ is $\frac{1}{2} g_\phi$, but still others use the same letter γ_ϕ for another way of expressing the coupling.)

The experimenters avoid all the ambiguity by noting that the diagram above implies that a free ϕ meson would have a certain rate to disintegrate into e^+e^-. They give this rate (which comes more directly from experiment anyway)

$$\Gamma\phi \to e^+e^- = 1.36 \pm .1 \times 10^{-6}\ \text{GeV}$$

either directly or as a branching ratio. A simple calculation shows the connection (for any vector meson V of mass m_V, neglecting electron mass) ($\alpha = 1/137$) is

$$\Gamma V \to e^+e^- = \frac{4\pi\alpha^2}{3} F_V^{\,2}\, m_V^{-3} = \frac{\alpha^2}{3}\,\frac{4\pi}{g_V^{\,2}}\, m_V \quad . \tag{14.3}$$

For the ϕ, $F_\phi = .080\ (\text{GeV})^2$ or $g_\phi^{\,2}/4\pi = 13.3$.

One can look in this energy region for other products (such as 3π, or $\eta^0\gamma$) and find the resonance again - thus determining the hadronic amplitudes for $\phi \to 3\pi$ or $\phi \to \eta^0\gamma$, usually given in the form of branching ratios.

In the case the final state is 3π there is another resonance at 785 MeV, the ω meson - it is studied in the same way.

Again studying 2π final states a large resonance near 765 of width around 125 MeV. This large width makes considerable ambiguity (in assumptions of how

Γ varies with q^2) in determining the constants, mass and width. Further the curve is asymmetric, sharper on the high energy side. A small shoulder is nearly apparent - this effect is interpreted as interference with the ω resonance assuming there is a finite amplitude for $\omega \to 2\pi$. (This violates isospin, and is an electromagnetic effect which we will discuss in the next lecture.)

Values for the various constants for the vector meson may be found in the Particle Properties Table. There is now some additional data from the Orsay Storage Rings (J. Lefrancois, 1971 International symposium on electron and photon interactions at high energy). Some of the differences from Particle Table results are new data but some are due to an altered way of reducing data, especially for the ρ. We shall have to wait for new particle tables to thrash these differences out.

B = Branching ratio

	New	Particle Tables
ϕ	$B(\phi \to \eta^o \gamma) = (2.1 \pm .7) \times 10^{-2}$	$B(\phi \to K^+ K^-) = 46.4 \pm 2.8\%$
	$B(\phi \to \pi^o \gamma) = (.25 \pm .09) \times 10^{-2}$	$B(\phi \to K_1 K_s) = 35.4 \pm 4\%$
	$B(\phi \to 3\pi) = (14.7 \pm 2.2) \times 10^{-2}$	$B(\phi \to 3\pi) = 18.2 \pm 5\%$
	$\Gamma\phi$ total $= 4.7 \pm .7$ MeV	$4.0 \pm .3$ MeV
ω	$B(\omega \to \pi^o \gamma) = .07 \pm .02$	$B(\omega \to \pi^o \gamma) = 9.3 \pm 1.2\%$
	$B(\omega \to \eta^o \gamma)$ Not seen $< .02$	$B(\omega \to 3\pi) = 90 \pm 4\%$
	$\Gamma\omega$ total $= 9.2 \pm 1.0$ MeV	$\Gamma\omega = 11.4 \pm 0.9$
	$B(\omega \to 2\pi) = .04 \pm .02$ (Phase $87^o \pm 15^o$)	$B(\omega \to 2\pi) = .009 \pm .002$
	$\Gamma(\omega \to e^+e^-)$ $.76 \pm .08$ keV	$B(\omega \to e^+e^-) = (.0066 \pm .0017)\%$
ρ	$m_\rho^2 = 780 \pm 6$	$m_\rho^2 = 765 \pm 10$
	Γ_ρ total $= 153 \pm 13$ MeV	Γ_ρ total $= 125 \pm 20$
	$\Gamma_\rho \to e^+e^- = 6.1 \pm .5$ keV	$\Gamma_\rho \to e^+e^- = 7.5 \pm .9$ keV

Calculating from the latest Orsay data, we get

$$g_\rho^2/4\pi = 2.27$$

$$g_\omega^2/4\pi = 18.3$$

$$g_\phi^2/4\pi = 13.3$$

but Orsay making "corrections for finite width" reduces the data in some other way:

$$g_\rho^2/4\pi = 2.56 \pm .22$$
$$g_\omega^2/4\pi = 19.2 \pm 2$$
$$g_\phi^2/4\pi = 11.3 \pm 0.8$$

Note. Now that we have data for the rates $\phi\to\eta\gamma$ and $\omega\to\pi\gamma$ we can compare the predictions of the quark model with harmonic dynamics of Feynman, Kislinger and Ravndal.

	Quark Model	Orsay
$\Gamma\phi\eta\gamma$	$= 1.79 \times 10^{-4}$ GeV	$1.0 \pm .3 + 10^{-4}$ GeV.
$\Gamma\omega\pi\gamma$	$= 1.92 \times 10^{-3}$ GeV	$.6 \pm .2 \times 10^{-3}$ GeV.

Lecture 15

We will make a digression from our main discussion of vector mesons to consider the interesting feature of $\omega - \rho$ interference.

The reason ω can go into 2π is that the state $|\omega\rangle$ is not pure $|\omega_o\rangle = \frac{1}{\sqrt{2}}(u\bar{u} + d\bar{d})$ isospin zero but has a small admixture of $|\rho_o\rangle = \frac{1}{\sqrt{2}}(u\bar{u} - d\bar{d})$ isospin one in it. This mixing is due to electrodynamics.

We consider the $\omega - \rho$ system as a two-state system where $C\rho_o$ is the amplitude to be in a ρ_o state and $C\omega_o$ is the amplitude to be in an ω_o state.

$$i\frac{d}{dt}\begin{pmatrix} C\omega_o \\ C\rho_o \end{pmatrix} = \begin{pmatrix} H\rho_o\rho_o & H\rho_o\omega_o \\ H\omega_o\rho_o & H\omega_o\omega_o \end{pmatrix}\begin{pmatrix} C\omega_o \\ C\rho_o \end{pmatrix} \tag{15.1}$$

the mass matrix is

$$H = \begin{pmatrix} m_\omega - \frac{i\Gamma\omega}{2} & \delta \\ \delta & m_\rho - \frac{i\Gamma\rho}{2} \end{pmatrix} = \begin{pmatrix} 784-i6 & \delta \\ \delta & 765-i62 \end{pmatrix}$$

therefore the true ω is

$$|\omega\rangle = |\omega_o\rangle + \frac{\delta}{m_\omega - m_\rho - \frac{i\Gamma\omega - i\Gamma\rho}{2}}|\rho_o\rangle$$

$$|\omega\rangle = |\omega_o\rangle + \frac{\delta}{19+56i}|\rho_o\rangle \tag{15.2}$$

δ can be given in terms of the branching ratio for $\omega\to 2\pi$

$$\frac{\text{Prob. }\omega\to 2\pi}{\text{Prob. }\omega\to\text{all}} = \left|\frac{\delta}{19+56i}\right|^2 \frac{\rho\to 2\pi}{\omega\to\text{all}} = 2.97\times 10^{-3}\ \delta^2\ \text{MeV}^{-2} \qquad (15.3)$$

From Orsay experimental data the above branching ratio is 4 ± 2% we therefore obtain $|\delta| = 3.7 \pm .9$ MeV. The phase of the ω-ρ interference is determined by fitting the "shoulder" in $\sigma\ (e^+e^- \to \pi^+\pi^-)$ near the ρ resonance; the value is 87° ± 15°. If δ is negative the phase of $\delta/(19 + 56i)$ is 109°, which is in fair agreement with the experiment.

We can understand the mixing as due to two effects: (1) The electromagnetic energy of $u\bar{u}$ is not equal to the energy of $d\bar{d}$ because of the difference in self energy of the objects and the difference of energy of interaction. (2) There is a contribution due to annihilation $\rho_o \to \gamma \to \rho_o$, $\omega_o \to \gamma \to \omega_o$ $\rho_o \to \gamma \to \omega_o$.

We can get estimates for the two above effects from the knowledge of $K^{*+} - K^{*o}$ and $\rho^+ - \rho^o$ mass differences and from F_ρ. Let the self energy of d be a and the mutual energy of $d\bar{d}$ be $-b$; then the self energy of u is 4a and the mutual energy of $u\bar{u}$ is $-$ 4b because the photons act twice on double the charge.

From $\omega_o = \frac{1}{\sqrt{2}}\ (u\bar{u} + d\bar{d})$, $\rho_o = \frac{1}{\sqrt{2}}\ (u\bar{u} - d\bar{d})$, $\rho^+ = u\bar{d}$, $K^{*+} = u\bar{s}$ and $K^{*o} = d\bar{s}$ we obtain the following electromagnetic self energies

$$\begin{aligned}
\rho^o &: = (4a + a) - \tfrac{1}{2}\,(4b + b) = 5a - 5b/2 \\
\omega^o &: = (4a + a) - \tfrac{1}{2}\,(4b + b) = 5a - 5b/2 \\
\rho^+ &: = (4a + a) + 2b \qquad\quad = 5a + 2b \\
K^{*+} &: = (4a + a) + 2b \qquad\quad = 5a + 2b \\
K^{*o} &\ \ = (a + a) \ \ - b \qquad\quad\ \ = 2a - b
\end{aligned} \qquad (15.4)$$

But we also have a matrix element between $\langle\rho^o|$ and $|\omega^o\rangle$

$$\langle\rho^o|\Delta m|\omega^o\rangle = (4a - a) - \tfrac{1}{2}\,(4b - b) = 3a - 3b/2$$

from this alone the mass matrix is

	ρ_o	ω_o
ρ_o	5a − 5b/2	3a − 3b/2
ω_o	3a − 3b/2	5a − 5b/2

Since $K^{*+} - K^{*o} = 3a + 3b$

$\rho^{+} - \rho^{o} = 9b/2$

we have $3a - 3b/2 = (K^{*+} - K^{*o}) - (\rho^{+} - \rho o')$ (15.5)

in terms of measurable quantities.

Note that in eq. (15.5) we have written $\rho^{o'}$ instead of ρ^{o}, this is only to indicate that $\rho^{o'}$ does not contain the contribution from the annihilation term which we calculate next.To first order in e^2 the change in mass is given by

$$\Delta m_{\rho}^{2} = \frac{4\pi e^{2}}{q^{2}} <\rho|J^{V}|o> <o|J^{V}|\rho> \qquad (15.6)$$

or

$$\Delta m_{\rho}^{2} = \frac{4\pi e^{2}}{m_{\rho}^{2}} F_{\rho}^{2} \qquad (15.7)$$

$$\Delta m_{\rho} = 1.53 \text{ MeV}$$

Let x be the amplitude for $\frac{1}{\sqrt{2}}(u\bar{u} - d\bar{d}) \rightarrow \gamma$, the amplitude for $\frac{1}{\sqrt{2}}(u\bar{u} + d\bar{d}) \rightarrow \gamma$ is x/3 (see lecture 16). The mass matrix due to the annihilation term is therefore proportional to

	ρ_o	ω_o
ρ_o	x^2	$x^2/3$
ω_o	$x^2/3$	$x^2/9$

(15.8)

We already have one entry of this matrix, $\Delta m_{\rho} = 1.53$ MeV we therefore have for the full matrix

	ρ_o	ω_o
ρ_o	1.53	.51
ω_o	.51	.17

MeV (15.9)

Adding the contributions from the electromagnetic self mass and from the annihilation term to the non-diagonal matrix element we find

$$\delta = .5 \text{ MeV} + (K^{*+} - K^{*0}) - (\rho^+ - \rho^{0'}) \tag{15.10}$$

and since

$$\begin{aligned} -(\rho^+ - \rho^{o'}) &= -(\rho^+ - \rho^{o}) - 1.53 \text{ MeV} \\ \delta &= -1.02 \text{ MeV} + (K^{*+} - K^{*o}) - (\rho^+ - \rho^{o}) \end{aligned} \tag{15.11}$$

from data in the particle tables, $(K^{*+} - K^{*0}) = -8 \pm 3$ MeV and $(\rho^+ - \rho^0) = -2.4 \pm 2.1$ MeV. But we do not trust these results, especially the ρ mass difference, we only say that they indicate that δ is likely to be negative.

The $\gamma\rho$ coupling of the form $\sqrt{4\pi e^2}\ F_\phi\ e_\mu^{\ \gamma} e_\mu^{\ \phi}$ is not gauge invariant and would give a finite mass^2 for the photon via

The coupling is $A_\mu^{\ \gamma} B_\mu^{\ \rho}$. However we can couple the fields $F_{\mu\nu}$ of the photon and ρ, for example. It leads to

$$\sqrt{4\pi e^2}\ C_\rho\ (q_\mu e_\nu^{\ \gamma} - q_\nu\ e_\mu^{\ \gamma})\ (q_\mu e_\nu^{\ \rho} - q_\nu e_\mu^{\ \rho}) \tag{15.12}$$

$$= \sqrt{4\pi e^2}\ \ 2\ C_\rho\ (q^2\ e^\gamma \cdot e^\rho - (q \cdot e^\gamma)\ (q \cdot e^\rho))$$

which is evidently gauge invariant. Since $e^\gamma \cdot q$ is always zero we see our pole behaves like

$$\frac{q^2(e^\gamma \cdot e^\phi)}{q^2 - m_\rho^2} = \frac{m_\rho^2}{q^2 - m_\rho^2}\ e^\gamma \cdot e^\rho + e^\gamma \cdot e^\rho \tag{15.13}$$

The latter term has no pole singularity and is "lost" in the background of other than pole term and effectively we can use a residue just proportional to $e^\gamma \cdot e^\rho$(i.e. $2\ C_\rho m_\rho^{\ 2} = F_\rho$).

Lecture 16

Vector mesons (continued)

We now return to our discussion of vector mesons. We can see what SU_3 gives for the ratios of the couplings $g_V^{\ 2}$ given at the end of lecture 14. We use the simple quark model (for counting only) and assuming the ϕ is purely strange quarks. J couples to $Q\bar{Q}$ pair proportional to its charge

$$\rho = \frac{1}{\sqrt{2}}(u\bar{u} - d\bar{d}) \rightarrow \frac{1}{\sqrt{2}}\left(\frac{2}{3} - \left(-\frac{1}{3}\right)\right) = \frac{1}{\sqrt{2}}$$

$$\omega = \frac{1}{\sqrt{2}}(u\bar{u} + d\bar{d}) \rightarrow \frac{1}{\sqrt{2}}\left(\frac{2}{3} + \left(-\frac{1}{3}\right)\right) = \frac{1}{3\sqrt{2}} \tag{16.1}$$

$$\phi = s\bar{s} \rightarrow -1/3 = -1/3$$

Noting that these are values for the coupling F, and that g is the reciprocal we find

$$g_\rho^{-2} : g_\omega^{-2} : g_\phi^{-2} = 9 : 1 : 2$$

Various people have tried to correct this for SU_3 breaking, but nobody really knows how. There are two questions. The first is, to what extent is the ϕ pure $s\bar{s}$? I see no way to determine that. The low value of the $\phi \rightarrow 2\pi$ branching ratio is interpreted in the quark model by saying that the ϕ being made of purely strange quarks finds it hard to go into non-strange objects (π). If so, all the amplitude to go to $\pi\gamma$ comes from an admixture of the ω state, $(u\bar{u} + d\bar{d})/\sqrt{2}$ in the ϕ wave function. In this way the amplitude to find ω in the ϕ comes out to about .10, although there is some uncertainty $\pm$ .03 as to how mass factors enter. This is only 1% probability, the state is 99% pure $s\bar{s}$.

Problem. Make a theory to estimate how much $\phi \rightarrow 3\pi$ would be expected if it is due to the fact that ϕ has a small admixture of ω state; assuming (for no excellent reason) that $s\bar{s}$ cannot go to 3π.

More important is the question of SU_3 breaking because the masses are not equal. For one example of ambiguities, should we compare F_V's directly with the SU_3 predicted ratios, or should it be F_V/m_V, or F_V/m_V^2 or what, which should bear the simple ratios? SU_3 cannot tell us and nobody really knows how to calculate it, although various guesses have been made. (I prefer F_V/m_V for a theoretical reason, but one which is really not profound or necessary). These questions probably do not strongly affect the ρ,ω ratio, as the masses are close together. As you see the ratio 9 : 1 is not very far off, (and, in spite of the uncertainties, the ϕ is also about right).

Vector Meson Dominance Model

To summarize we exhibit the photon to $\pi^+\pi^-$ amplitude near the pole of the

ρ (the only one of the three ϕ, ω, ρ which couples to the pion) as

$$\langle\pi^+\pi^-|J_\mu|o\rangle \approx \langle\pi^+\pi^-|\rho\rangle \frac{1}{q^2-m_\rho^2+i\Gamma_\rho m_\rho} \langle\rho|J_\mu|o\rangle$$

$$\approx f_{\rho\pi\pi}\,(p_{1\mu}+p_{2\mu})\,\frac{1}{q^2-m_\rho^2+i\Gamma_\rho m_\rho}\,F_\rho \tag{16.2}$$

$$\approx (f_{\rho\pi\pi}/g_\rho)\,\frac{m_\rho^2}{q^2-m_\rho^2+i\Gamma_\rho m_\rho}\,(p_{1\mu}+p_{2\mu})$$

when $q^2 \approx m_\rho^2$. One would ordinarily expect other terms (such as a direct coupling of photon to pions or various other "intermediate states") to be added to this. This, of course, would not change the behaviour near the pole (which, unfortunately for the ρ is not so definite as the width $\Gamma\rho$ is rather large, which makes many difficulties in practice in fitting it, nevertheless in this theoretical discussion we shall neglect it). Another way to say it is to suppose the numerator has a factor $\xi(q^2)$ which varies with q^2 such that $\xi(m_\rho^2) = 1$, or again to say the "constants" $f_{\rho\pi\pi}$ or g_ρ vary with q^2 only their value at $q^2 = m_\rho^2$ being defined. All these ways say the same thing so we shall not argue about them.

Nevertheless a bold hypothesis has been suggested (vector meson dominance) that this expression with $f_{\rho\pi\pi}/g_\rho$ constant is all there is to it - there are no other terms. I see no good physical reason for this. It is made in analogy to another mysterious hopothesis of the same kind, PCAC, that works. There we do an analogous thing with a pion pole - but only extend it from $q^2 = m_\pi^2 = .02$ at the pole, to $q^2 = 0$. Here we suggest eq. (16.2) is valid not only near $q^2 = m_\rho^2$ but also for all q^2 or at least to $q^2 = 0$.

We know for long wave lengths the pion looks like a simple point charge so that as $q^2 \to 0$ (use crossing to convert $\langle\pi^+\pi^-|J_\mu|o\rangle$ to $\langle\pi^+|J_\mu|\pi^+\rangle$ we have

$$\langle\pi^+|J_\mu|\pi^+\rangle = (p_{1\mu}+p_{2\mu}) \text{ for } q^2 \to 0, \tag{16.3}$$

hence for (16.2) to be true we must have (neglect problems generated by $\Gamma\rho$, or suppose $\Gamma\rho$ depends on q^2 and vanishes at the $(2m_\pi)^2$ threshold and below).

$$f_{\rho\pi\pi}/g_\rho = 1$$

Experimentally $f^2_{\rho\pi\pi}/4\pi = 2.43$

$$g_\rho^2/4\pi = 2.56$$

well within the 10% uncertainties in determining these quantities.

It would also imply that the form factor $F_\pi|q^2|$ for the pion, written $<\pi^+|J_\mu|\pi^+> = F_\pi(q^2)(p_{1\mu} + p_{2\mu})$ would, for negative q^2, have to be exactly $(1+(-q^2)/m_\rho^2)^{-1}$. We shall look at what evidence there is later - so far it looks very good.

The idea is extended to the case of interactions with any state - for example in interactions with the nucleon the part due to isovector would be

$$<N|J_\mu|N> \text{ isovector} = -\frac{m_\rho^2}{g_\rho} \frac{1}{(q^2-m_\rho^2)} <N|\rho|N> \qquad (16.4)$$

ρ with polarization μ. The term $<N|\rho|N>$, the coupling of a ρ to a nucleon is defined only at the ρ pole, and there has an electric and magnetic part so we can write as $<N|C_1\gamma_\mu + C_2 \frac{1}{2}(\gamma_\mu \not{q} - \not{q}\gamma_\mu)|N>$. Thus, expressing the isovector part of the current coupling in terms of the usually defined form factors we have $<N|J_\mu|N> = <N|F_1^V(q^2)\gamma_\mu + F_2^V(q^2)\frac{1}{2}(\gamma_\mu \not{q} - \not{q}\gamma_\mu)|N>$

$$= \frac{1}{(1-q^2/m_\rho^2)} <N|\frac{C_1}{g_\rho}\gamma_\mu + \frac{C_2}{g_\rho}\frac{1}{2}(\gamma_\mu \not{q} - \not{q}\gamma_\mu)|N> \qquad (16.5)$$

Thus we predict that $F_1/F_2 = C_1/C_2$ is independent of q^2 (fair) and F_1 varies as $1/(1-q^2/m_\rho^2)$. This latter is not good, $1/(1-q^2/m_\rho^2)^2$ is better. Here we are testing (1) for large negative q^2 - far from the ρ pole and it fails. Obviously VMD cannot be an exact principle for all q^2. But maybe it is nearly valid from $q^2 = m_\rho^2$ to 0. We must discuss extensions of the idea and further evidence.

A further conclusion from the nucleon case above is that $C_1/g_\rho = 1$. There is no way to get the ρ coupling to the nucleon, C_1 or C_2, without ambiguity at present (or ever? is it a matter of definition $<N|\rho|N>$ cannot have all the particles physically on the mass shell!). Some knowledge, from πN scattering for example, exists for the ccupling of the ρ-like trajectory - or for what we call ρ-exchange - but the extrapolation from ρ-trajectory to ρ-pole exchange is ambiguous. The latter (ρ-pole exchange) has an amplitude varying as s but the terms associated with ρ trajectory exchange do not, and they are not supposed to in Regge theory itself.

A similar expression using the ω trajectory is to give the isoscalar coupling. Since the mass of the ω is nearly equal to that of the ρ, the isoscalar and isovector form factors should have the same dependence as q^2 - they do even though this dependence is not the $(1-q^2 m_\rho^2)^{-1}$ expected.

Lecture 17

Vector Meson Dominance Model (continued)

In greater generality all three neutral vector mesons may be involved. For example, in coupling to an arbitrary hadron state we write

$$\langle \text{Hadrons}|J_\mu|0\rangle = \sum_{V=\rho,\omega,\phi} \frac{1}{g_V}\left(\frac{-m_V^2}{q^2-m_V^2+i\Gamma_V m_V}\right)\langle \text{Hadrons}|V\rangle \tag{17.1}$$

V with polarization μ

(or of course by crossing for any transition hadrons X → hadrons Y)

$$\langle X|J_\mu|Y\rangle = \sum_V \frac{1}{g_V}\left(\frac{-m_V^2}{q^2-m_V^2+i\Gamma_V m_V}\right)\langle XV_\mu|Y\rangle \tag{17.2}$$

That the left side of (17.2) gives the total of all the ways that J_μ can couple is the general statement of the hypothesis of vector dominance. We have already seen how it predicted the $\rho\pi\pi$ coupling constant. Can we do a similar thing for the ϕKK coupling, say ϕK^+K^-? We know that the current operator at $q^2 \approx 0$ might be written $(p_{1\mu}+p_{2\mu})$ with coefficient 1 because the charge is 1, so we have the sum of $\frac{1}{g_V}$ $(K^+K^-|V)$ must equal 1. Thus if we write

$$\langle K^+K^-|V\rangle = f_{VK^+K^-}(p_{1\mu}+p_{2\mu}) \tag{17.3}$$

we would have

$$1 = \sum_V f_{VK^+K^-}/g_V \tag{17.4}$$

The problem now is that although we do know the three g_V (from experiment) we have too many unknowns in (17.4) to relate it to experiment. (The rate that $\phi \to K^+K^-$ implies $f_{VK^+K^-}/4\pi = 1.47$). We need some rule to partition the bits of current between ρ, ω, and ϕ.

A very suggestive thing to try is to express the quark model idea that the ϕ is pure $s\bar{s}$, and thus couples only with the s quarks in a system. This suggests we invent three currents J^u, J^d, J^s which couple only to the u quarks, the d quarks, or the s quarks (as though each had unit charge). Then, for example, the electric current (charges on u, d, s, are + 2/3, -1/3, -1/3 respectively) is:

$$J^{el} = +\frac{2}{3} J^u - \frac{1}{3} J^d - \frac{1}{3} J^s \tag{17.5}$$

Then, the ρ being $\frac{1}{\sqrt{2}}$ $(u\bar{u} - d\bar{d})$, we define a "ρ-type current" as $J^\rho = \frac{1}{\sqrt{2}} (J^u - J^d)$, likewise an ω-type $J^\omega = \frac{1}{\sqrt{2}} (J^u + J^d)$ and a ϕ type, which since $\phi = \bar{s}s$ is just $J^\phi = J^s$ hence for example

$$J^{el} = \frac{1}{\sqrt{2}} J^\rho + \frac{1}{3\sqrt{2}} J^\omega - \frac{1}{3} J^\phi \ . \tag{17.6}$$

Now obviously we suppose the three pieces into which J^{el} is split in (17.6) are the corresponding vector meson resonant pieces of (17.5). This may be written in a very simple way, if the g_v really have the ratios of the quark model (Lecture 14) we are saying, of course

$$\langle Y|J_\mu^V|X\rangle = \text{const.} \left(\frac{-m_v^2}{q^2-m_v^2+i\Gamma_v m_v}\right) \langle YV_\mu|X\rangle \tag{17.7}$$

where the constant, the same for all quarks is $\sqrt{2}/g_\rho$.

However we still must define in an experimentally definite way what J^u, J^d, J^s, or J^ρ, J^ω, J^ϕ are. We can use isospin, hypercharge and quark number (or rather, baryon numbers) as three conserved quantities (each definite for a particular transition $X \to Y$) to serve instead of quark numbers. Thus let J^Z represent the current of Z-component isospin; J^Y current of hypercharge, J^B current of baryon number (equals 1/3 times the current of quark number) to write

$$\begin{aligned} J^U &= J^Z + \frac{1}{2} J^Y + J^B \\ J^d &= -J^Z + \frac{1}{2} J^Y + J^B \\ J^s &= - J^Y + J^B \end{aligned} \tag{17.8}$$

(because if you substitute for example the quantum numbers of the u quark, $I_Z = + 1/2$, $Y = + 1/3$, $B = 1/3$ for the currents you get one for J^u, zero for J^d and J^s, etc.)

Hence

$$J^\rho = \sqrt{2} J^Z$$

$$J^{\omega} = \sqrt{2}\left(\frac{1}{2}J^{Y} + J^{B}\right) \tag{17.9}$$

$$J^{\phi} = -J^{Y} + J^{B} \quad .$$

This then defines precisely what we mean by the currents in (14.5) and (14.6).

For example, let us apply this to the K^+K^- decay of the ϕ. We need the couplings of J^{ϕ} to K^+ (we also calculate J^{ρ}, J^{ω} for completeness). For the K^+, $I_Z = +1/2$, $Y = +1$, $B = 0$, so at zero q^2;

$$\langle K^+|J^{\rho}|K^+\rangle = 1/\sqrt{2}$$

$$\langle K^+|J^{\omega}|K^+\rangle = 1/\sqrt{2} \tag{17.10}$$

$$\langle K^+|J^{\phi}|K^+\rangle = -1$$

Thus combining this with (14.9) and (14.7) we see we must have

$$f_{\rho K^+K^-}/g_{\rho} = \frac{1}{2} \; ; \quad f_{\omega K^+K^-}/g_{\omega} = \frac{1}{6} \; ; \quad f_{\phi K^+K^-}/g_{\phi} = \frac{1}{3}$$

defining how the total of (1) in (17.4) is partitioned. At any rate we predict

$$(f_{\phi\, K^+K^-}^{2}/4\pi)/(g_{\phi}^{2}/4\pi) = 1/9$$

which agrees perfectly with the $f_{\phi\, KK}^{2}/4\pi = 1.47$ from $\phi \to K^+K^-$ and $g_{\phi}^{2}/4\pi = 13.3$ from $\phi \to e^+e^-$.

How slow is $\phi \to 3\pi$ compared to $\omega \to 3\pi$? ϕ goes 18% to 3π, but Γ is only 4 MeV so the partial width is only .7 MeV while it is 10 MeV for the ω. Phase space is larger for the ϕ but more detailed analysis would require knowledge of how the matrix element varies. If e is the polarization of the vector meson and P_1, P_2, P_3 the four vectors of the three pions the amplitude must be constant $e_{\mu\nu\sigma\rho}\, e_{\mu}\, P_{1\nu}\, P_{2\sigma}\, P_{3\rho}$. The "constant" for the ϕ has to be about 0.1 of its value for the ω.

Lecture 18

ϕ as $\bar{s}s$

The ϕ appears to act as pure $\bar{s}s$. What is the significance of this - or rather how can we define this in terms of quantum numbers or rules without referring to the quark model? We don't know how. We try to say that it couples weakly to states which have no strange particles in them, so not to 3π but yes to $\bar{K}K$. Yet such an idea "a state that has no strange particles in it" is not readily definable. Due to strong interactions a state like $\bar{K}K$ has no overall quantum numbers to distinguish it from 3π; in fact via virtual interactions the $\bar{K}K$ should couple strongly to 3π (for example through a virtual ω, $\bar{K}K \leftrightarrow \omega \leftrightarrow 3\pi$.) Therefore the question remains, what keeps the ϕ from coupling strongly with 3π?

From the point of view of a quark model we might try to say that the $s\bar{s}$ state is selected from the $\frac{1}{\sqrt{2}}(u\bar{u}+d\bar{d})$ state of the ω simply by the fact that s quarks carry a larger mass (or a different interaction energy) than non-strange quarks. So in lowest order of perturbation theory the eigenstates of energy are $s\bar{s}$ and $\frac{1}{\sqrt{2}}(u\bar{u}+d\bar{d})$. But then we find these states are unstable, and as a kind of perturbation they decay, it turns out with small widths. But this "perturbation" is not small - there is a coupling of ϕ^o and ω^o to every meson state (with the ω coupling equal to $-\frac{1}{\sqrt{2}}$ of the ϕ coupling). Thus for these virtual meson states for example $\phi \to K\bar{K} \to \omega$ there are diagrams

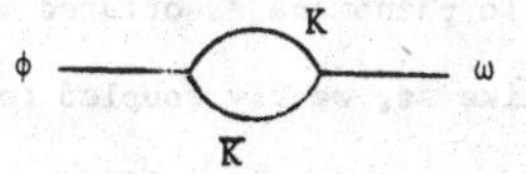

which mix the ϕ, ω state. These diagrams cannot be calculated, (the above diverges quadratically) and no calculation of such virtual strong interactions in any problem has ever been successful. But usually the quantitative idea that if the states can be connected via strongly interacting virtual states they can go into each other is valid. The coupling constants $f_{\phi K\bar{K}}$ and $f_{\omega K\bar{K}}$ are so large they would seem to have effects of order 1, that is, to strongly mix the original pure $s\bar{s}$ and $\frac{1}{\sqrt{2}}(u\bar{u}+d\bar{d})$.

Of course it is always possible that the world is simply complicated, that the combination $s\bar{s}$ is not isolated by principle in the beginning; but the very virtual interactions select some linear combination of $s\bar{s}$ and $\frac{1}{\sqrt{2}}(u\bar{u}+d\bar{d})$ to be the ϕ (and the orthogonal combination for the ω) and it just comes out that the combination is $s\bar{s}$.

Then perhaps it is really possible to understand why $\phi \to \pi\gamma$ is so small, but is the smallness of $\phi \to 3\pi$ really obvious in this model of the state?

But still more striking is the fact that this "accident" is repeated again!

For the mesons of spin 2^+ where the f(1260) goes nicely into 2π or 4π and into $K\bar{K}$. The f(1514) goes predominately into $K\bar{K}$ and even into $K\bar{K}^* + \bar{K}K^*$ and not into $\pi\pi$ in spite of the even larger phase space. (The quark model rates (Feynman, Kislinger, Ravndal) using pure $\frac{1}{\sqrt{2}}(u\bar{u}+d\bar{d})$ for the f(1260) and $s\bar{s}$ for f(1514) are all high, but the relative proportions are right.)

The experimental and quark model (FKR) rates for 2^+ mesons are:

		Experimental	FKR
f(1260) →	$\pi\pi$	120 MeV	244 MeV
	$K\bar{K}$	∿ 8	13
	$2\pi^+2\pi^-$	10	
f(1514) →	$K\bar{K}$	52	103
	$\bar{K}K^* + K\bar{K}^*$	7	15
	$\pi\pi$	< 10	0
	$\eta\pi\pi$	13 ± 7	
	$\eta\eta$	< 30	

NOTE: There are in addition two hadronic phenomena associated with the ϕ which are unexpected unless the ϕ is, like $s\bar{s}$, weakly coupled to hadronic states of zero strangeness. The first is backward production of ϕ in $\pi + N \to N + \phi$

It is very weak, as though the ϕ were not coupled to the N, N* (a nucleon trajectory) junction. Again $p + \bar{p} \to \phi + n\pi$ is very strongly suppressed (relative to $\rho + n\pi$ for example) as though the ϕ's s and $\bar{s}$ cannot be made from the non-strange quarks in the other particles.

<u>VDM and Photon Hadron Interactions</u>

Complete vector dominance implies that the photon cannot interact with a hadron except as it first becomes a ρ, ω or ϕ, and then this ρ, ω or ϕ (say V in general) interacts with the hadrons. Thus we would expect to find some kind of relation between amplitudes like

$$\text{Amp}\ (\gamma + A \to B) = \sqrt{4\pi e^2}\ \sum_V \frac{1}{g_V}\ (V + A \to B). \tag{18.1}$$

There are several questions involved in using the formula. First, there are theoretical questions of meaning. Since γ has $q^2 = 0$, the vector meson on the right side should be a virtual vector meson of $q^2 = 0$, for which strictly no definition can be given. It might be argued that since mass on the initial state incoming particle might have the least effect at high s, (where $E \approx p$ anyway for finite mass) this relation may be most nearly correct for real vector mesons on the right-hand side at very high s. Another related ambiguity is that the photon has only two polarizations, helicity ± 1, so the equation has only meaning for the corresponding helicity ± 1 states of V. But helicity is not a relativistically invariant concept and depends on the frame. Most theorists have come to the conclusion it is in the s-channel frame that the ρ helicity should be ± 1. At any rate this uncertainty is reduced by going to large s also. Thus we shall, to avoid much complicated discussion, limit ourselves to comparisons to experiment at the highest energies now available.

The next question is how we are to obtain the vector meson cross sections or amplitudes $V + A \to B$, after all V meson beams are not available. Sometimes theoretical arguments are available, but if they are too complicated they are not useful to test VDM. The most useful simple cases are :

a) Pseudoscalar meson production, in particular $\rho + N \to \pi + N$.

b) Diffraction (elastic) scattering from nucleons $\rho + N \to \rho + N$ or nucleus ρ + Nucleus $\to \rho$ + Nucleus.

We discuss each in turn.

We can study the reaction $\rho + N \to \pi + N$ by experimentally studying the reversed reaction $\pi + N \to \rho + N$, which should have the same amplitude. This has been studied at 15 GeV and reported by D.W.G.S. Leith, Phenomenology Conference, 1971 (Caltech) p. 555. Naturally the ρ is not observed directly but $\pi^- p \to \rho^0 n$ is inferred from a complete study of $\pi^- p \to \pi^+ \pi^- n$ looking in the appropriate mass region of the two outgoing pions. Some corrections must be made for pairs of pions, say in mutual s waves, which are not due to virtual ρ decay. However the data at low t has been nicely analyzed in detail theoretically and we can describe the results, and compare to VDM expectations.

We evidently wish to compare these results to reactions like $\gamma + p \to \pi^+ + n$ and $\gamma N \to \pi N$ is related directly by VDM to $\pi^- p \to V^0 n$ where V^0 is a vector meson linear combination of ρ, ω, ϕ. Since the ϕ coupling to mesons is small we drop it, and the ω ϕ interference. VDM then predicts for example that (by squaring appropriate amplitudes)

$$\frac{1}{2}\left[\frac{d\sigma}{dt}(\gamma p\to\pi^+ n) + \frac{d\sigma}{dt}(\gamma n\to\pi^- p)\right]$$

$$= 4\pi e^2\left[\frac{1}{g_\rho^2}[\rho_{11}]_\rho \frac{d\sigma}{dt}(\pi^- p\to\rho^0 n) + \frac{1}{g_\omega^2}[\rho_{11}]_\omega \frac{d\sigma}{dt}(\pi^- p\to\omega n)\right] \quad (18.2)$$

These $[\rho_{11}]$ are density matrix elements to project out the helicity $\pm$ 1 reactions. Data in the ω reaction are lacking, but $g_\rho^{-2}/g_\omega^{-2}$ is about 9/1 so the second term is probably never large. In addition, at 8 GeV $\frac{d\sigma}{dt}$ (π^- p $\to\rho^0$ n) is itself about 10 times $\frac{d\sigma}{dt}$ (π^- p $\to\omega$ n) in the forward direction, so the ω contribution might be as small as 1% to the VDM result. The same could not have been said of a ρ-ω interference term (for an amplitude of order .1 interfering with 1 can make a 20% effect, but its square is only 1%). That is the reason that the comparison is made to the sum of $\gamma p\to\pi^+$ n and $\gamma n\to\pi^-$p cross sections - for in that sum the ρ-ω interference term (from isospin considerations) cancels out (as does the ρ-ϕ interference).

The cross sections for all the helicity combinations of the ρ have been measured (and reported in the form of density matrices) by measuring the angular distributions of the pair of pions $\pi^+\pi^-$ resulting from the disintegration of the polarized ρ^2. We also have data with polarized photons (perpendicular and parallel to the plane of production and so can make two comparisons. The vector dominance model predicts then,

$$\frac{d\sigma_\perp}{dt}(\gamma N\to\pi N) = \frac{4\pi e^2}{g_\rho^2}[\rho_{11}+\rho_{1-1}]\frac{d\sigma}{dt}(\pi^- p\to\rho^0 n) \quad (18.3)$$

$$\frac{d\sigma_{//}}{dt}(\gamma N\to\pi N) = \frac{4\pi e^2}{g_\rho^2}[\rho_{11}-\rho_{1-1}]\frac{d\sigma}{dt}(\pi^- p\to\rho^0 m) \quad (18.4)$$

where $d\sigma/dt$ ($\gamma N\to\pi N$) means the average of $d\sigma/dt$ ($\gamma n\to\pi^-$p) and $d\sigma/dt$ ($\gamma p\to\pi^+$n) and the ρ's are density matrix elements in the helicity frame. The results of such a comparison appear as follows (polarization data for photons at 15 GeV are not yet available, but a guessed extrapolation from 8 GeV was made - this does not affect the comparison of the sum of $\sigma_\perp$ and $\sigma_{//}$ which is of course measured as the unpolarized cross section at 15 GeV).

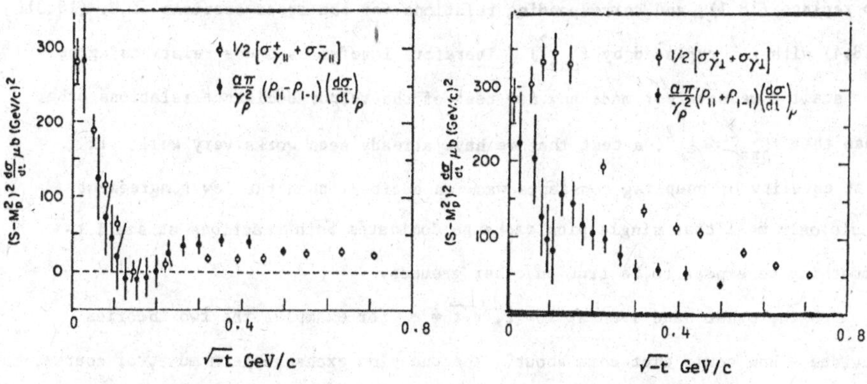

The results are seen to agree in exactly the forward direction and generally for the unnatural parity exchange $\sigma_{//}$, but clearly disagree away from $t = 0$ for $\sigma_{\perp}$. Thus VDM as a general theorem cannot be exactly correct (barring the possibility of an enormous ω contribution in this region).

Thus photons do not couple exactly as off mass shell vector mesons. But just how do they couple? To make progress we should study the nature of the deviations from a strict VDM and find their characteristics. For example had we a good theory, which at present we do not have - but should be able to develop - we could say why these various cross sections agree for $\sigma_{//}$ and for the forward direction, and why not for $\sigma_{\perp}$? Perhaps some clues can be obtained from this example so we study it further.

The values of t we are studying here are so small that single pion exchange ought to dominate both cross sections (i.e. γ as well as ρ). If this were exactly true without absorption corrections we would expect a relation between the cross sections similar to (18.2), (18.3), or (18.4) simply because diagrams correspond:

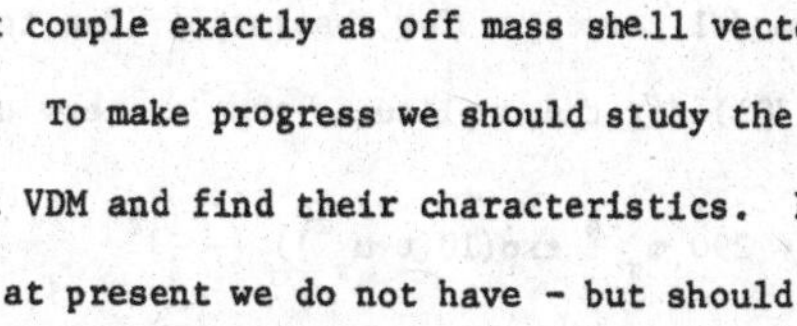

The only difference (aside from the mass^2 difference of ρ, γ and selection of the correct helicity amplitudes to compare) would be that the γ is coupled to the $p_{1\mu} + p_{2\mu}$ of the pions via a factor $\sqrt{4\pi e^2}$ and the ρ via $f_{\rho\pi\pi}$. Hence we would expect a relation like

$$\text{Amp}\ (\gamma N \to \pi N) = \sqrt{4\pi e^2}\ \frac{1}{f_{\rho\pi\pi}}\ \text{Amp}\ (\rho N \to \pi N) \tag{18.5}$$

to replace (18.1), and corresponding relations for the cross sections (i.e. (18.3), (18.4) with g_ρ^2 replaced by $f_{\rho\pi\pi}^2$). Therefore insofar as these relations agree for small t we have not made any new test of the vector dominance relations other than that $f_{\rho\pi\pi}^2 \approx g_\rho^2$, a test that we have already seen works very well. If that equality of coupling constants was an accident then the low t agreement would only mean that single pion exchange dominates both reactions at small t - something we expect to be true on other grounds.

On the other hand even at low t, $\sqrt{-t} = m_\pi$ for example, the two theories deviate - how could that come about? The one pion exchange term must, of course, be corrected for absorption. The differences between $\sigma_{\gamma N}$ and $\sigma_{\rho N}$ at low t most likely can be attributed to differences in the degree to which absorption modifies the one pion exchange expectation.

In the first place (see Leith's report) a fair fit to the ρ data is given by a one pion exchange model corrected for absorption (due to P.K. Williams, Phys. Rev. D1 1812 (1970)) for the amplitudes that concern us this gives

$$s^2[\rho_{11} + \rho_{1-1}]\frac{d\sigma_\perp}{dt} = 290\ m_\pi^{-4}\ \exp(10(t-m_\pi^2))\ [-\tfrac{1}{2}]^2 \tag{18.6}$$

and

$$s^2[\rho_{11} - \rho_{1-1}]\frac{d\sigma_{//}}{dt} = 290\ m_\pi^{-4}\ \exp(10(t-m_\pi^2))\ [\frac{t}{t-m_\pi^2} - \tfrac{1}{2}]^2 \tag{18.7}$$

We have already seen that the unabsorbed one pion exchange gives for these amplitudes (in connection with the discussion of photoproduction)

$$A_\perp \sim 0$$

$$A_{||} \sim \frac{t}{t-m_\pi^2}$$

and that ideal absorption would subtract 1/2 from each amplitude. This leads to zero $A_{//}$ if $-t = m_\pi^2$ and for large t equal amplitudes for $A_\perp$, $A_{//}$ and hence to an asymmetry that rises from 0 at t = 0 to 1 at $t = m_\pi^2$ and falls beyond that to 0. This agrees with the asymmetry observed for the ρ, but not for the photon. In the photon case the fall off for large t is only moderate, to say 0.5 or so. Thus for the photon it looks like a more appropriate correction for absorption might be to subtract more like 2/3 from the two amplitudes. (Thus the asymmetry is 1 for $t = 2m_\pi^2$ and falls to 0.6 for large t - not in disagreement with experiment). We really do not exactly know why the subtracted term for absorption should always be the full 1/2. Possibly in this case extra contributions (isoscalar or isovector?) are contributing to the photon case to increase the 1/2 appropriate to the ρ case. The point here is an effect different for ρ and γ and hence different from expectations of VDM (unless it could possibly arise from the ω term). However the correct constant term to use is determined experimentally by the cross section for $t = 0$ which is just proportional to the square of this constant with known coefficients. This determines that the constant is indeed 1/2 to 10%.

But a much more striking difference is the difference of $\sigma_{\gamma\perp}$ and $\sigma_{\rho\perp}$. This difference is already large at small t - the $\sigma_{\rho\perp}$ drops off much faster than does $\sigma_{\gamma\perp}$. Why? I don't know. I think the mystery here is why the $\sigma_{\rho\perp}$ falls so fast. In the fit to OPE plus absorption it was found empirically that a factor $e^{10(t-m_\pi^2)}$ needed to be included in these cross sections. P. K. Williams had suggested such a factor for absorption effects but expected a much slower fall off (like $e^{3(t-m_\pi^2)}$). To what is such a rapid fall off due? Such a rapid variation is entirely unexpected. It is very possibly due to the method of analyzing the data. The effect of an amplitude to produce a pair of π's in a mutual s-wave (not a ρ) has to be subtracted away. The value of $\rho_{//}$ near $-t = m_\pi^2$ is especially sensitive to what is done here (at $t = 0$ or $-t \sim 10m_\pi^2$ it is not sensitive.) Nevertheless there is a contribution to natural parity exchange which is different for γ and ρ (violating VDM). How can we make a theory of where and when such deviations should arise? We leave it as a problem to analyze this in more detail. One obvious possibility seeing the large number in the exponent is to be reminded that total absorption effects (as seen in elastic scattering) do fall off as expbt with b of order 8 or 9 for π nucleon scattering. So perhaps a careful analysis of absorption, or also of the possibility that the source of the pion is indefinite by the size of the nucleon or the ρ, will explain it. The point of explaining it, however, is that the drop off is much slower for the photon case. Therefore whatever the cause it works differently in ρ and γ pseudoscalar production. In its study lies the possibility of understanding physically where the ideas of VDM go wrong. The problem cannot be difficult - the effect shows up for small t and hence for large impact parameter and therefore in a realm where physical phenomena are usually understandable.

Lecture 19

Diffractive Production of ρ, ω, ϕ

The next topic we take up is the diffractive production of vector mesons by photons. At first we study the ρ meson for more detailed data is available here. I will not go into as much detail in the results as we are accustomed to - for a full recent report see:

Wolf: 1971 International Conference on Electron and Photon Interactions at High Energy, Cornell, Ithaca, N.Y., (1971).

Our main concern will be a comparison to VDM.

The ρ^o production in $\gamma N \rightarrow \rho^o N$ looks very much like diffraction scattering, the cross section approaches a constant at high energy and has the typical dependence of such scattering. But how does the photon diffract into a ρ? One answer is provided by VDM, from the expected amplitude relation (18.1) we expect

$$\frac{d\sigma}{dt}(\gamma N \rightarrow \rho^o N) = \frac{4\pi e^2}{g_\rho^2}\frac{d\sigma}{dt}(\rho^o N \rightarrow \rho^o N) \qquad (19.1)$$

(because of I spin change it is expected that $\omega^o N \rightarrow \rho^o N$ or $\phi^o N \rightarrow \rho^o N$ will fall rapidly with energy so only the ρ term remains). We do not know directly the cross section of ρ^o on nucleons but we can expect it to be a typical elastic scattering.

A crude use of the quark model supposing the quarks to scatter independently of their spin direction and to be similarly distributed inside the ρ^o and π^o suggests $\sigma(\rho^o N) = \sigma(\pi^o N)$. This latter is not known by direct experiment but isospin gives it as $\frac{1}{2}[\sigma(\pi^- N \rightarrow \pi^- N) + \sigma(\pi^+ N \rightarrow \pi^+ N)]$. It does turn out that the total cross section for $\gamma N \rightarrow \rho^o N$ varies with energy in just the way $\frac{1}{2}[\sigma(\pi^+ N) + \sigma(\pi^- N)]$ behaves. In fact the total cross section is given correctly by this rule and equation (19.1) with $g_\rho^2 = 2.8$.

There is a 15% uncertainty because the observed process is $\gamma N \rightarrow N + \pi^+ + \pi^-$ and there are uncertainties in interpretation due to so-called Deck-type diagrams in which the γ becomes two pions not at the ρ resonance and one pion scatters like

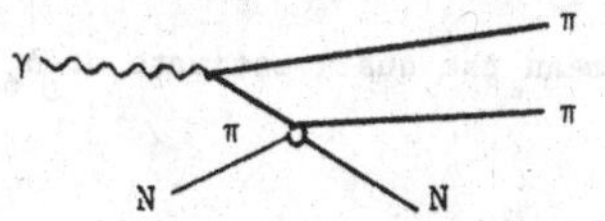

instead of

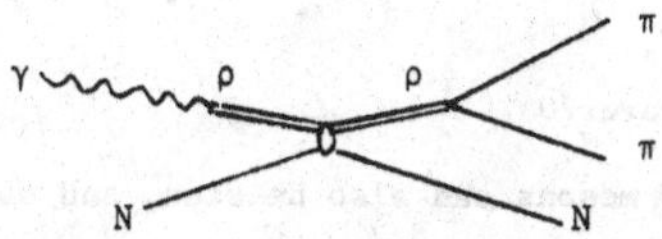

A strict adherence to VDM would, as far as I can see expect the analogue for ρ scattering to 2π via

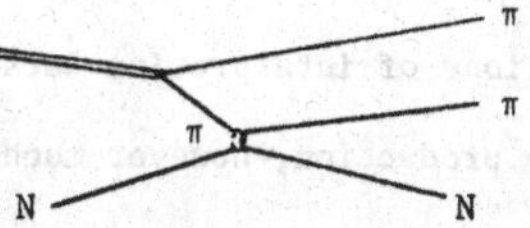

but the point is in estimating just the effect of the simple ρ elastic scattering so we can use our π^o scattering analogy.

This idea that $\gamma N \to \rho N$ tells us about $\rho N \to \rho N$ permits us to summarize the behavior in another way. The polarization of the ρ coming out can be measured. It is very nearly purely transverse and polarized as the incident γ ray. This is excellent evidence that the process of elastic scattering (also called pomeron exchange) does not change the helicity in the center of mass system.

ω production. Here the diffraction term is smaller so other processes are also effective, in particular one pion exchange at the lower energies. The natural parity exchange has besides the pomeron exchange also possible A_2 exchange so the purely diffractive part has not been separated out yet for a clear test.

ϕ production. The experiments show some inconsistency but data is available.

One way to test the relations of VDM is to compare the forward differential cross section for $\gamma N \to VN$ with the forward differential cross section for $VN \to VN$. This latter can probably be estimated fairly well from the optical theoreom using quark model estimates for the total cross sections for VN. (The quark model estimate for the ρ, ω and ϕ are:

$$\sigma_T(\rho p) = \sigma_T(\omega p) = \frac{1}{2}\,[\sigma_T(\pi^+ p) + \sigma_T(\pi^- p)] = 28 \text{ mb at 5 GeV}$$

$$\sigma_T(\phi p) = \sigma_T(K^+ p) + \sigma_T(K^- p) - \sigma_T(\pi^+ p) = 15 \text{ mb at 5 GeV.)}$$

This gives a fit with $g_\rho^2/4\pi = 2.6 \pm .3$, $g_\omega^2 = 24 \pm 5$ and $g_\phi^2 = 22 \pm 6$. Only the g_ϕ^2 seems too big. It could mean the quark estimate of $\sigma_{\phi N}$ is too big - 10 mb would fit better.

Lecture 20

Diffraction production of vector mesons can also be seen, and more copiously and clearly from γ on nuclei, A. (Reference: K. Gottfried "Nuclear Photoprocesses and vector dominance", Cornell Conference, 1971.) There are difficulties for the ω since the final 3π state is hard to measure, and the ρ is so wide that theoretical questions of interpreting data are involved. The clearest case experimentally is ϕ production, however much more data exists for the ρ. In these experiments on ρ production observations are made of 2π production by photons on nuclei, of course. A beautiful resonance in mass of the 2π is seen at the ρ mass. It is asymmetrical showing the interference effect with $\omega \to 2\pi$ just as in e^+e^- production.

The ϕ data is now good. The dependence on A, the mass number of the nucleus, gives us some idea of $\sigma_{\phi p}$, and the absolute cross section permits a determination of g_ϕ. However an uncertainty arises for the results depend sensitively on the choice of an unknown quantity, the real part of the forward ϕN scattering amplitude $f_{\phi\phi}$. Let $\alpha_{\phi N} = \mathrm{Im} f_{\phi\phi}/\mathrm{Re}\, f_{\phi\phi}$. Then the data is insufficient to determine all three quantities $g_\phi^2/4\pi$, σ_ϕ, and $\alpha_{\phi N}$. If $g_\phi^2/4\pi$ is its Orsay value 13.3 then we can only conclude that α may be perhaps from -.3 to -.5 and $\sigma_{\phi N}$ in the range 8 to 14 mb (at about 7 GeV), possibly a bit lower than the quark model rules would suggest, but little can be said.

The ρ data is more extensive. Here the large width causes some confusion as to what part of the data is to be attributed to ρ production. Again we have three parameters $g_\rho^2/4\pi$, $\sigma_{\rho N}$ and $\alpha_{\rho N}$. One gets agreement with expectations if we choose $\alpha_{\rho N} = -0.24$ (but it is not well determined). The data gives the "good values" (at 8.8 GeV) $g_\rho^2/4\pi = 2.6$ and $\sigma_{\rho N} = 27$ mb. (But $\alpha_{\rho N}$ is not well determined and results depend on it.) Therefore the phenomenon of vector meson production exists and behaves closely like a diffraction process, and could well be in quantitative agreement with the VDM. Does this test the model? I think not because formula (19.1) applied to diffraction scattering can be derived (at very high energy) from another hypothesis - namely that the elastic scattering of the ρ is much larger than diffractive dissociation scattering of the ρ (the process by which $\rho N \to \rho^* N$ where ρ^* is some other state of the same isospin as the ρ, and which survives as $s \to \infty$). Such diffraction dissociation (as $NN \to N^*N$ or $\pi N \to \pi^* N$) are in other reactions perhaps only 30% of the elastic scattering from the nucleon so there is no reason not to assume it for the ρ. Furthermore for the ρ we shall only have to assume that the part of diffraction dissociation to those ρ^* having spin 1^- (like the photon) is small compared to the elastic. There is also a direct experimental confirmation of the fact that $\gamma N \to \rho^* N$ where ρ^* is any isospin one, 1^- state, produced in a diffraction dissociative way does not fall with energy.

For production on nuclei our assumption (of the dominance of elastic diffraction dissociation) is ever more true as A rises - for the elastic comes from diffraction from the shadow of the entire nucleus - whereas the particles produced by dissociation can come only from its edge.

The point is it is not dominance of the γ to ρ that we require for the validity of equation (19.1), it would also follow from dominance of $\rho \to \rho$ in the products of the diffraction of ρ on nucleon and nucleus.

We can see how this works most clearly in the case of scattering from a large nucleus where our assumption is most nearly valid. Instead of considering $\gamma A \to \rho A$ analyze the reverse reaction $\rho A \to \gamma A$ and compare it to $\rho A \to \rho A$. The high energy ρA scattering appears, in say the c.m. as follows:

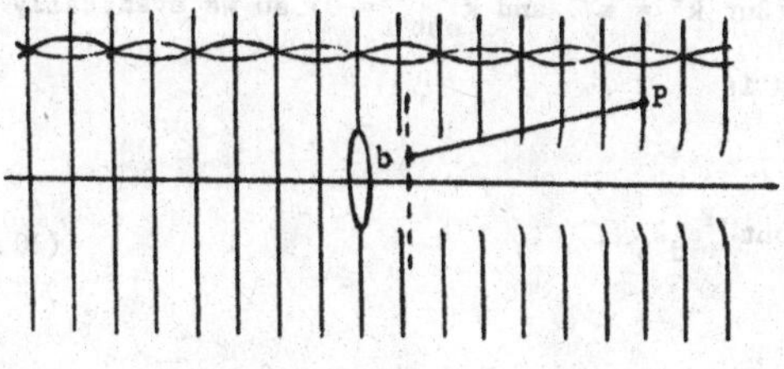

The waves of the incoming ρ come in from the left and fall upon the nucleus where they are absorbed. A short distance (which can be many wave lengths if $s \to \infty$ beyond the target say at the dotted line the wave function is nearly the original ρ wave function with a hole as a function of b, the impact parameter. Beyond that the waves slowly diffract into the shadow in a way determined by Huygens principle. If $1 - a(b)$ is the amplitude to find a ρ at b (so a is near 1 for b smaller than the nuclear radius, and near 0 beyond) the amplitude for an outgoing wave of transverse momentum Q is $\int(1-a(b))\, e^{iQ\cdot b}\, d^2b = f(Q)$. The amplitude at P is

$$\psi(P) = \int(1-a(b))\, \frac{e^{ikr}}{r}\, d^2b \quad \text{approximately} \tag{20.1}$$

Of course the 1 gives only the forward beam and doesn't interest us - it is just as if the ρ came out only across the face of the nucleus. To be sure, if the ρ were not a point particle but say made of parts (indicated by the braided line) we could not precisely define everything in terms of an amplitude to give a ρ, but would need a function of all the parts as variables. But clearly for b outside the nucleus the parts are in the same relative amplitude as in a ρ - only near the edge is there some distortion of the relative behavior of the parts - and so a projection possible into some other than the ρ. But the amplitude a(b) is clearly that for a ρ except near the edge; so that elastic scattering is much larger (for nuclei, at least) than diffraction dissociation.

Now we have noted from experiment by e^+e^- that whenever we have a ρ present there is an amplitude that this ρ will disintegrate to e^+e^- (via a photon) and therefore there is a source of the electromagnetic field - a current given by $j_\mu = \sqrt{4\pi e^2}\, F_\rho \times$amp. to find ρ polarized in direction μ, ($F_\rho = -M_\rho^{\,2}/g_\rho$).

Thus in our problem at each point P there is a current source of electromagnetic field of size $\psi(P)$ - thus the total amplitude for finding a photon going out with wave vector $\vec{k}_{out}$ is $\int\psi(P)\, e^{-ik_{out}\cdot P}\, d^3P$. If we substitute our expression for $\psi(P)$ in this the convolution of e^{-ikr}/r and $e^{-ik_{out}\cdot P}$ just gives $e^{-ik_{out}\cdot b}\, \dfrac{1}{k^2-k_{out}^{\,2}}$. (But $k^2 = m_\rho^{\,2}$ and $k_{out}^{\,2} = 0$) so we eventually get the outgoing amplitude for photon is

$$\frac{\sqrt{4\pi e^2}\,F_\rho}{m_\rho^{\,2}} \int (1-a(b))e^{-iQ_{out}\cdot P}\, d^2b \tag{20.2}$$

where Q_{out} is the transverse part of k_{out}. The integral is the same as for the ρ elastic scattering so the amplitudes are proportional - the cross sections must bear the ratio $4\pi e^2/g_\rho^{\,2}$.

One can of course carry the mathematics out in more detail separating the z and transverse momentum carefully to see just how the assumption of high energy is valid. It is easy to understand if the diffraction of the wave into the shadow is omitted. Then the wave going forward behind the shadow is just $e^{ik_z^\rho z}(1-a(b))$ where the incident ρ is of energy E and $k_z^\rho = \sqrt{E^2-m_\rho^2} = E - m_\rho^2/2E$. The amplitude to make a photon of energy E, transverse momentum Q, hence longitudinal momentum $k_z^\gamma = E - Q^2/2E$ as

$$\frac{\sqrt{4\pi e^2}\,F_\rho}{2E}\int e^{i(k_z^\rho-k_z^\gamma)z}\,(1-a(b))\,e^{iQ\cdot b}\,d^2b =$$

$$\frac{\sqrt{4\pi e^2}\,F_\rho}{i2E(k_z^\rho-k_z^\gamma)}\,f(Q) = \frac{\sqrt{4\pi e^2}\,F_\rho}{m_\rho^2-Q^2}\,f(Q) \qquad (20.3)$$

A 1/2E factor has been included to take account of the relativistic $1/\sqrt{2E}$ normalizations for the incoming and outgoing particles. The $m_\rho^2 - Q^2$ is wrong here, the effect of diffraction into the shadow makes it m_ρ^2 (as though the k_z^ρ were $E - \frac{m_\rho^2+Q^2}{2E}$ for ρ's moving in this direction).

It is evident that the polarization of the γ is the same as that of the ρ for this process. This feature has been checked experimentally very carefully and is very striking, the ρ polarization produced by polarized photons is nearly purely that of the photon.

Often it has been suggested that VDM predicts its formula (19.1) via a diagram like

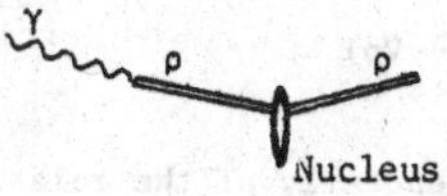

suggesting the photon becomes a virtual ρ of zero q^2, and hence that the coupling F_ρ (or $-m_\rho^2/g_\rho$) should be that appropriate for $q^2 = 0$ instead of $q^2 = m_\rho^2$ where it is measured in e^+e^-. It is a test of VDM to find that this constant is unchanged in this range. But we have seen, by consideration of the reverse reaction, neglecting diffraction dissociation that the appropriate constant should be that of m_ρ^2 and no extrapolation is involved.

J. Mandula has confirmed my argument by an argument from dispersion theory and the reduction formula.

The assumption of neglect of diffraction dissociation comes in here. Near the edge, the parts of the ρ may not be in precisely the same relative motion to produce a ρ, but they might still produce a photon. Clearly this is small. It means an interfering term of order $F_{\rho*}/m_{\rho*}^2$ times the amplitude to make $\rho*$ where ($F_{\rho*}$ is the coupling of $\rho*$ to photons) summed on $\rho*$. For nuclei at least it must be small coming only near the edge. The point is not so much that $F_{\rho*}$ is small (which it may be, as so little $\rho*$ seems to be produced by photons on nuclei (which measures $F_{\rho*}^2\ \sigma_{\rho*N}$) but that the number of $\rho*$ produced by ρ via diffraction dissociation is small.

Other Tests of VDM

If VDM were correct we expect the photons to have amplitudes $1/g_V$ to be various vector mesons so that the total cross section of γ's on p is $1/g_V^2$ times the total cross section of each of the vector mesons. We expect

$$\sigma_{tot}(\gamma p) = \sum_{\rho,\omega,\phi} \frac{4\pi e^2}{g_V^2}\ \sigma_{tot}(Vp) \tag{20.4}$$

(neglecting possible interference effects of ω and ϕ). This is dominated by the ρ (because of the 9:1:2 ratio of g_V^2 and $\sigma(\phi p)$ is small). We estimate $\sigma_t(\rho p) = \sigma_t(\omega p)$ by the quark model as equal $\sigma_t(\pi^0 p)$ (a scheme which we checked in our $\sigma(\gamma p \to Vp)$ considerations above). We find experimentally that $\sigma_{tot}(\gamma p)$ is too large by 40%, as though other processes were available to the photon to interact with the proton besides going through virtual vector mesons.

A very similar check giving precisely the same result is to use eq. (18.1) to say

$$A(\gamma p \to \gamma p) = \sqrt{4\pi e^2}\ \sum_V \frac{1}{g_V}\ A(\gamma p \to Vp) \tag{20.5}$$

Again the dominant term is the ρ. Independently of the relative phases we get the inequality

$$d\sigma_0/dt\ (\gamma p \to \gamma p) \le 4\pi e^2 \left| \sum_{\rho,\omega,\phi} \sqrt{\frac{1}{g_V^2}\ \frac{d\sigma}{dt}\ (\gamma p \to Vp)} \right|^2 \tag{20.6}$$

The quantities on the right hand side are available directly by experiment. The dependence of $d\sigma_0/dt$ ($\gamma p \to Vp$) on s (from 2.7 to 5.2 GeV) and on t (from 0 to .4 GeV^2) are well represented by the right hand side, but it is always only about 1/2 of the experimental cross section! Thus VDM fails here.

This is in accord with the 40% error in the total cross section text, for the optical theoreom relates $d\sigma/dt$ ($\gamma p \to \gamma p$) in the forward direction to the square of $\sigma_{tot}(\gamma p)$ (assuming something about phases).

Lecture 21

Shadowing in Nuclei

How should the total cross section for γ+ Nucleus vary with the mass number of the nucleus A? We know for collisions with hadrons nuclear matter is nearly opaque (because σ hadron-nucleon is comparable to the spacing of the nucleons) and therefore the cross section for A nucleons is not the sum of a contribution from each (hence $\sigma_{tot} \sim A\,\sigma$ nucleon) because nucleons in the front shadow those in back so ultimately for large A it goes as the area of the nucleus or $A^{2/3}$.

On the other hand for photon-nucleus collisions a first glance would suggest that the nucleus now being transparent ($\sigma_{\gamma\,\text{nucleon}}$ is much smaller than nuclear spacing in a nucleus) each nucleon would see the full beam and hence the cross section would go as A (times the simple nucleon photon cross section).

On the other hand the VDM shows that this latter conclusion cannot be right in general for it is incorrect if the VDM is correct. A photon amplitude should be proportional to the ρ amplitude - the latter is a hadronic amplitude - hence the photon cross section proportional to the ρ cross section and therefore varying as $A^{2/3}$ (I neglect the contributions of ω and ϕ in this qualitative discussion, their contributions are easily reinstated).

The reason is that in the simple view we imagined the interaction γ + nucleon = X to be a local process of γ on nucleon, the γ interacting near where the nucleon is located - but VDM reminds us that this is not so. A γ can become virtual hadrons (e.g. 2π, but most importantly a ρ) far away in front of the target, and the virtual hadrons propagate a long way to the nucleon to interact. A real photon is a pure ideal photon plus a virtual hadron with amplitude $1/\Delta E$ according to perturbation theory where ΔE is the energy difference of the state of given momentum as photon and as hadron. Thus if it is a ρ the energy is $\sqrt{\nu^2+m_\rho^2}$ while that of the photon is ν so $\Delta E = \sqrt{\nu^2+m_\rho^2} - \nu \approx m_\rho^2/2\nu$, very small for large ν. This is also the distance ahead of the nucleon where the photon-hadron conversion occurs. If this is large compared to the mean free path of ρ in the nucleus then shadowing occurs ($A^{2/3}$), if it is small, no shadowing (A).

The physical idea can be seen best first for a simple model consisting of two thin slabs a, b of nuclear matter one in front of the other by d. We shall calculate according to VDM.

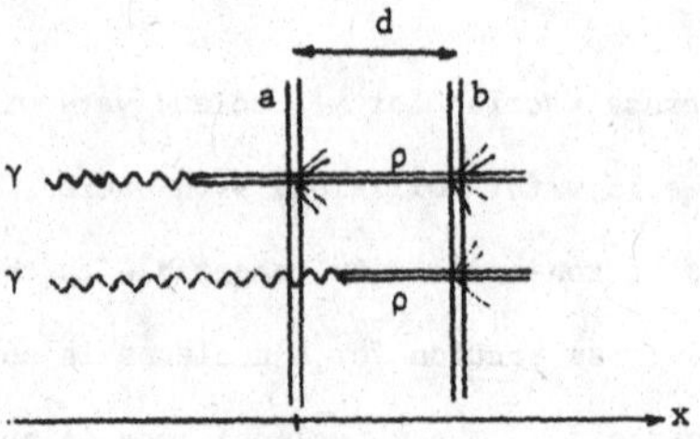

If a ρ were impinging suppose the small probability of being absorbed in the first layer is f_a; in b alone if ρ impinged f_b. But the probability that the ρ gets to b is $(1-f_a)$ so the product made in b is $(1-f_a)\,f_b$. The total cross section is thus $f_a+(1-f_a)f_b = f_a+f_b-f_af_b$ the last term representing shadowing.

Let us calculate the probability for any products made in b if a is present by photons via VDM. If a is not present the amplitude to find a ρ at b making product is proportional to the amplitude the γ converts to ρ at x ($\sqrt{4\pi e^2}\,F_\gamma$) and then the ρ arrives at d. This is

$$\sqrt{4\pi e^2}\,F_\rho\,e^{i(k_\gamma-k_\rho)(x-d)} \tag{21.1}$$

(the phase has been taken relative to the photon phase at x = d i.e. the above expression contains an additional factor exp $(-ik_\gamma d)$) where the k_γ, k_ρ are the k-vector for γ or ρ at the same frequency. Hence $k_\gamma - k_\rho = m_\rho^2/2\nu$ for large ν ($=k_\gamma$).

When a is present (21.1) is valid only for $d > x > 0$. For $x < 0$ we have the additional factor for ρ to get through a $= \sqrt{1-f_a} \approx 1-f_a/2$. Thus for $x < 0$ we have

$$\sqrt{4\pi e^2}\,F_\rho(1-f_a/2)\,e^{i(k_\gamma-k_\rho)(x-d)} \tag{21.2}$$

The total amplitude (integrate x) is therefore proportional to

$$\frac{\sqrt{4\pi e^2}\,F_\rho}{k_\gamma-k_\rho}\left(1-\frac{f_a}{2}\,e^{-i(k_\gamma-k_\rho)d}\right) \tag{21.3}$$

The main term has an amplitude proportional to $\dfrac{\sqrt{4\pi e^2}\,F_\rho}{m_\rho^2} = \dfrac{\sqrt{4\pi e^2}}{g_\rho}$ as we expect, the term in shadowing f_a comes in various phases - and in a more complete problem with a continuum of layers would cancel out for $(k_\gamma - k_\rho)d > 1$.

Thus the criterion for scattering involves ν, shadowing should be complete for $\nu \to \infty$

One can extend this easily to a thick layer in one dimension. We seek the amplitude to find a ρ at x (for that will later be used by squaring and summing x to get the total cross section.

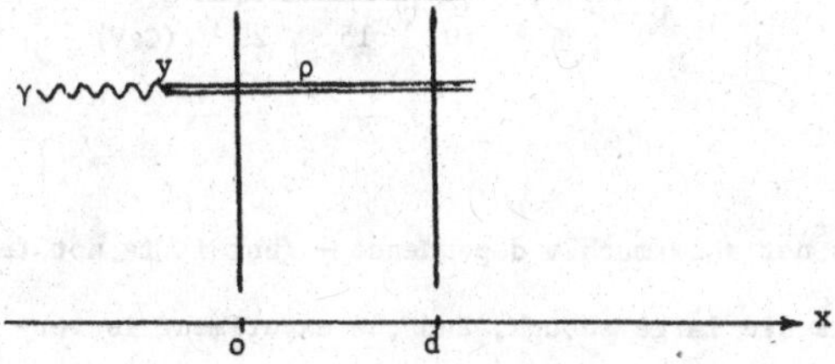

We have two cases $y < 0$ and $y > 0$

$y < 0$: γ converts at y, propagates as ρ to 0; propagates as ρ in medium to x

$y > 0$: γ converts at y, propagates in medium to x.

The amplitude is proportional to

$$e^{i(k_\gamma - k_\rho)y}\, e^{-i(k_\gamma - k_\rho')x} \qquad y < 0$$

$$e^{i(k_\gamma - k_\rho')y}\, e^{-i(k_\gamma - k_\rho')x} \qquad y > 0 \tag{21.4}$$

(the phases have been taken relative to the photon phase at x, i.e. the above expressions contain the additional factot $\exp(-ik_\gamma x)$).

Integrating over y the amplitude at x is

$$\frac{1}{k_\gamma - k_\rho}\, e^{-i(k_\gamma - k_\rho')x} + \frac{1}{k_\gamma - k_\rho'}\left(1 - e^{-i(k_\gamma - k_\rho')x}\right) \tag{21.5}$$

Here k_ρ' is k_ρ in the medium. At very high ν it has an imaginary part (representing absorption) which is finite and fixed, the real part goes, of course with ν, but may differ from it by a finite amount. Thus as $\nu \to \infty$, $k_\gamma - k_\rho' \approx k$ a fixed number (whose imaginary part gives the ρ absorption cross section and whose phase is the phase of forward ρ nucleon scattering).

Thus if $k_\gamma - k_\rho = \frac{m^2}{2\nu} \ll k$, the first term in (21.5) dominates and we get exact proportion to ρ absorption, hence absorption.

A full calculation including the spherical geometry of the nucleus has been made. They show a ν dependence of $\sigma_{tot}(A)/A\sigma(\text{nucleon})$ for various nuclei (see Gottfreid's report).

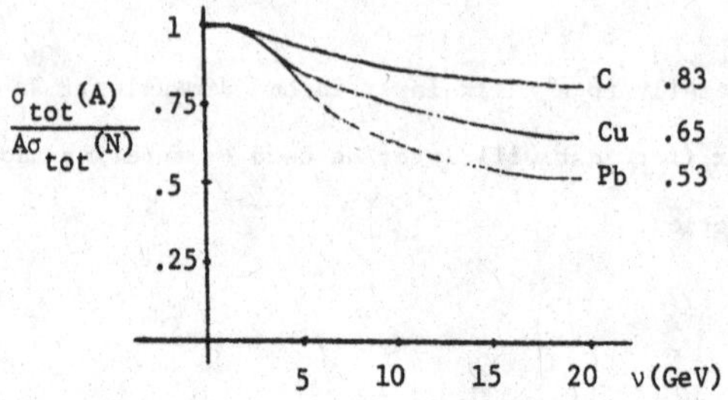

The data does not show much ν dependence - (but it is not far off, experimental errors are large enough, and the experiment is very hard at lower energies) but not outside expectations - however the asymptotic values of $\sigma_{tot}(A)/A\sigma_{tot(N)}$ are closer to .87 for C, .75 for Cu and .60 (with large error) for Pb.

These are all understandable if the total cross section is considered 70% VDM and 30% something else; (as we have seen is true) and if the something else goes just as A purely, showing no shadowing. Then the asymptotic (for $\nu \to \infty$) values of $\sigma_{tot}(A)/A\sigma(N)$ should differ from 1 by only 70% of the VDM theoretical value for this difference.

Measurements also exist for virtual photons. For them $q^2 = -Q^2$ and $k_\gamma = \nu - \frac{Q^2}{2\nu}$ so $k_\gamma - k_\rho = \frac{m_\rho^2 + Q^2}{\nu}$. Thus we expect from VDM that we again get a ν dependent shadowing, the parameter determining the extent of shadowing being $(m^2 + Q^2)/2\nu$ instead of $m^2/2\nu$. No ν dependence of shadowing is seen (in Au for $Q^2 = 0.5$) in strong disagreement with VDM predictions. If we correct VDM by allowing a certain percentage of "other processes" which do not show shadowing, of course, less shadowing is expected. But we should also note the fraction of something else must rise with Q^2. The ρ amplitude falls as $1/(m^2 + Q^2)$ so VDM cross sections should fall as $(m^2 + Q^2)^{-2}$. The actual cross section probably falls only as Q^{-2} for large Q^2 (see later) hence the fraction of photon cross sections which can be associated with virtual vector meson production (VDM contribution) must fall as photons become more virtual. Thus the results for Au are not surprising.

To Summarize the Position of VDM

Interaction of photons of small q^2 (or $q^2 = 0$) is very strongly influenced by the nearby poles at the mass of the vector mesons. Substantial parts of the interaction can be understood as being due to intermediate virtual vector meson coupling - but, of course, not all the coupling can be so described (for example only 70% of the total γp cross section at high energy is so described).

In the special case of the coupling to pion the dominance is more complete, as evidenced by direct experiment $e^+e^- \to \pi^+\pi^-$ (showing mainly the ρ, ω peak; not a large background and not much else resonant has been found to $q^2 = 4\text{GeV}^2$ and by the fit of constants $f_{\rho\pi\pi} \approx g_\rho$. (This will be discussed further in lecture 24, in reference to the dispersion theory of the pion form factor.)

Finally measurements exist on nuclei for photon production of incoherent ρ_o($-t > 0.1$ GeV) as well as for production of other particles such as π^+, π^- and even π^o (also K^+). Here the theory is complicated by a number of features - such as the absorption of the particle leaving the nucleus. The VDM theory predicts rapid variations in A_{eff} with energy, whereas experiments indicate only a small variation, if any. (See Gottfried, Ithaca Conference (1971) and Diebold, High Energy Physics Conference at Boulder, April 1969). There is at present no explanation known for this discrepancy - its study would make a nice problem. I doubt that the fact that the γ couples only partially to VDM is sufficient to account for the large discrepancy. Some item in the theory of these incoherent final states is probably involved.

Electromagnetic Form Factors

Lecture 22

Electromagnetic Form Factors

Nucleon: The matrix element for the interaction of an electron, say, and a proton involves the diagonal element of the current operator $\langle p|J_\mu|p\rangle$. Using relativistic invariance and gauge invariance this can be written in the form

$$\langle p_2|J_\mu|p_1\rangle = \langle \bar{u}_2|\gamma_\mu F_1 + \frac{1}{2}(\gamma_\mu \not{q} - \not{q}\gamma_\mu)F_2|u_1\rangle$$

where u_1 u_2 are 4-spinors describing the in and out proton of momenta p_1, p_2 momentum transfer $q = p_2 - p_1$. F_1 and F_2 are functions of q^2 only. Other linear combinations of F_1 and F_2 have also been defined, most used are (the so-called electric and magnetic form factors)

$$G_E = F_1 + (q^2/2M)F_2$$

$$G_M = F_1 + 2MF_2$$

clearly as $q^2 \to 0$ F_1 is the charge: $F_{1\ \text{proton}}(0) = 1$, $F_{1\ \text{neutron}}(0) = 0$ and F_2 are the anomalous magnetic momenta. Hence $G_E(0) = 1$ for p, 0 for n $G_M(0) = \mu_p$ is the total magnetic moment.

These have been measured by scattering electrons from protons elastically. The first order amplitude is ($\bar{u}_4$, u_3 are electron spinors)

$$(\bar{u}_4 \gamma_\mu u_3) \frac{4\pi e^2}{q^2} \langle p_2 | J_\mu | p_1 \rangle \tag{22.1}$$

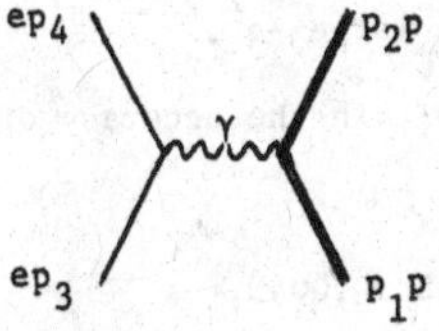

The cross section arising from this is:

<u>In lab</u>

$$\sigma(\theta) = \sigma_{NS}(\theta) \left[\frac{G_E^2 + \tau G_M^2}{1+\tau} + 2\tau G_M^2 \tan^2(\theta/2) \right] \tag{22.2}$$

(Rosenbluth)

$$\sigma_{NS}(\theta) = \frac{e^4 \cos^2 \theta/2}{4E_o^2 \sin^4 \theta/2 (1 + \frac{2E_o}{M} \sin^2 \theta/2)} \; : \; \tau = -q^2/4M^2 \tag{22.3}$$

$$-q^2 = \frac{4E_o^2 \sin^2 \theta/2}{1 + \frac{2E_o}{M} \sin^2 \theta/2}$$

By varying both E_o and θ we can vary q^2 and θ independently, and thus verify whether the bracket in (22.2) does vary linearly with $\tan^2(\theta/2)$ for fixed q^2. It works well, so both G_E and G_M can be extracted for the proton. Data for G_M is more accurate than G_E. For the neutron uncertainties of the deuteron wave functions make the determinations, especially of G_{En} much less certain.

Higher electromagnetic corrections do come in, especially effects from the bremstrahlung diagram

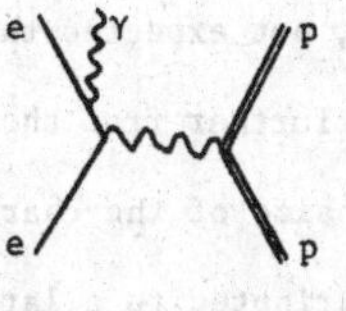

but they are largely understood and allowed for. This is an important but tedious correction to the data to get the ideal photon exchange cross section.

Experimental results are usually described by giving deviations from a set of completely empirical approximations, namely that

$$G_{Ep} = \frac{G_{Mp}}{\mu_p} = \frac{G_{Mn}}{\mu_n} = \left(1 + \frac{-q^2}{.71} \right)^{-2}$$

The latter function is called the dipole function G_D.

Most accurately known is G_{Mp}/μ_p. It vaguely follows the function G_D.

For small q^2 it varies as about $1 + (2.6 \pm .1)(+q^2)$, very close to the dipole. $G_{Mp}/\mu_p G_D \sim 1.08$ for $1 < -q^2 < 6$. Above $-q^2 = 8\ \mathrm{GeV}^2$ the ratio falls below 1. Thus G_M falls somewhat faster than $1/q^4$ for large $-q^2$.

The equality $G_{Mn}/\mu_n = G_{Mp}/\mu_p$ as within the accuracy of the neutron data.

For the neutron $G_{EM}{}^2$ at $q^2 = 1$ is $.0045 \pm .0043$.

One can fit with $G_{En} \approx G_{Mn}(\tau/(1 + 10\,\tau))$.

Neutron-electron scattering at low energy (using electrons in atoms) gives about

$$\left.\frac{dG_{En}\ (q^2)}{d\ q^2}\right|_{q^2 = 0} = 0.9\ \mu_n/4M^2$$

For the proton $G_{Ep}/(G_{Mp}/\mu_p)$ falls for larger q^2 near 2 or 3 so if put as $1 + A\,q^2$, A is perhaps .8 (large error). It is not possible that $G_{Ep} = G_{Mp}/\mu_p$ for all q^2, for at $q^2 = 4M^2$ the definition of these two G's in terms of F_1 and F_2 become identical, so unless G_{Ep} and G_{Mp} vanish here (which is very unlikely) we must have $G_{Ep} = G_{Mp}$ at $q^2 = 4M^2$. Thus $G_{Ep}/(G_{Mp}/\mu_p) = 2.79$ at $q^2 = +4M_p{}^2$ and $=1$ at $q^2 = 0$.

The simple ratios among G_{Mp}, G_{Mn}, G_{Ep}, G_{En} which are nearly right would be expected by pure vector meson dominance (assuming the mass of ρ and ω equal, and that ϕ couples only weakly to the proton). However, all the form factors would be multiples not of G_D but of the meson propagator $M_\rho{}^2/(M_\rho{}^2 - q^2) = (1 + Q^2/.58)^{-1}$. This supposes that the photon couples to a point charge nucleon via the ρ. Thus it has the wrong large Q^2 behavior Q^{-2}. Its small Q^2 behavior is $1 - 1.7\,Q^2$ instead of $1 - 2.6\,Q^2$. It is, of course, not expected that VDM continue to work as we go to negative Q^2 further and further from the pole at $Q^2 = -m_\rho{}^2$. But there is little doubt that the large size of the charge square radius (i.e. the coefficient -2.6 of Q^2) is contributed to a large extent by virtual ρ's. It is not really clear why these ratios do maintain their values at $Q^2 = 0$ out to $Q^2 = 2$ or $3\ \mathrm{GeV}^2$ as they do to 10 or 20%.

Lecture 23

Electromagnetic Form Factors (continued)

We can gain some insight into the meaning of form factors by looking at the non-relativistic case. For a spinless system the form factor is just the momentum representation of the charge density.

$$f(Q) = \int \rho(\bar{R})\, e^{iQ\cdot\bar{R}}\, d^3\bar{R} = \int \rho(r)\, 4\pi r^2 dr \left(\frac{\sin Qr}{Qr}\right) \tag{23.1}$$

Expanding in powers of Q^2 for small Q^2, with $\rho(r)$ normalized so that

$\int \rho(r)\, 4\pi r^2 dr = 1$ we get $(\sin Qr/Qr = 1 - Q^2 r^2/6)$

$$f(Q) = 1 - \frac{1}{6} r_p^{\;2}\, \omega^2 + \ldots \tag{23.2}$$

where $r_p^{\;2}$ is the mean square charge radius

$$r_p^{\;2} = \int r^2\, \rho(r)\, 4\pi r^2 dr$$

An exponential $e^{-\alpha r}$ for the charge density gives the dipole distribution $f(Q) = (1 + Q^2/\alpha^2)^{-2}$. The Q^{-4} dependence says the slope at origin must be finite not zero. It is, however, not a valid conclusion to use non-relativistic views to study large Q^2. Nevertheless the conclusion is essentially correct.

Relativistically the fact that as $Q^2 \to \infty$ the form factors go to zero means that there is zero probability to find the proton alone as a point charge. This could result in field theory either from there being no ideal field entity (parton) with quantum numbers of the proton, or else, if there were such the real proton is always dressed, it has zero qmplitude of being an ideal proton.

To make a relativistic analogue of a thing like the hydrogen atom (whose charge distribution is $e^{-\alpha r}$) we imagine two spinless particles held together with a spinless Yukawa potential generated by the weak exchange of some scalar meson, then (1) the non-relativistic wave function shows a $e^{-\alpha r}$ behavior, (2) the form factor varies as $1/Q^4$ for large Q^2.

For example, let one particle A be charged of mass m_1, the other B neutral of mass m_2 held weakly non relativistically. The mass of the proton is near $m_A + m_B$ and neglecting in zeroth order the relative motion the incoming proton of momentum p_1, means A has momentum ap_1 and B has momentum bp_1 $(a = \frac{M_A}{M_A+M_B}, a+b=1)$. Finally we want a bound proton of momentum p_2 hence parts of momentum ap_2, bp_2. We do this by hitting A with the very large $Q = p_1 - p_2$, whereas for the proton ultimately to hold together p_A must change only by aQ, hence an amount of momentum bQ must be passed over to b (by an exchange of the meson responsible for interaction of A and B. The diagrams are

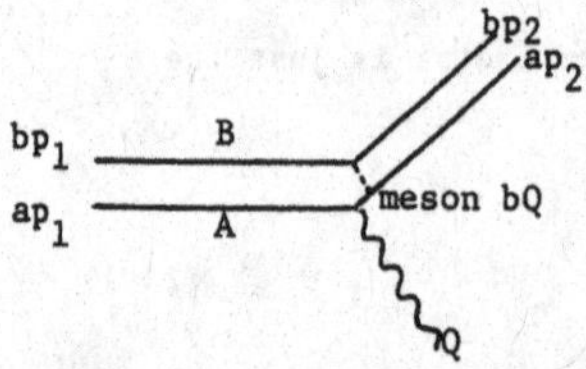

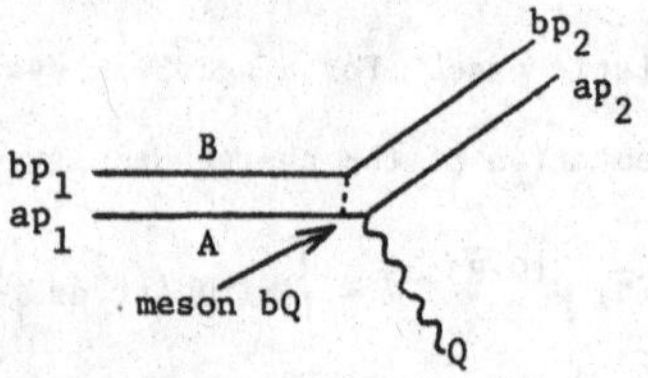

The two propagators each give $1/Q^2$ for large Q^2 hence $1/Q^4$ for the form factor. Thus the first diagram is

$$\frac{1}{(ap_1+Q)^2}\ \frac{1}{(bQ)^2} = \frac{a}{b^3Q^4}$$

the second

$$\frac{1}{(ap_2-Q)^2}\ \frac{1}{(bQ)^2} = \frac{a}{bQ^4}$$

at large Q^2.

If vector mesons are exchanged between a particle of spin 1/2 and one of spin zero, a Q^2 appears in the numerator due to magnetic coupling.

Non relativistically if a system is made of 3 particles, the large Q^2 asymptotic behavior depends not on the singularity when just two come together, but rather when all three are on top of one another.

As we shall see, such pictures are too simple and inadequate. We are only making very naive models to compare to the non-relativistic ideas. They will not agree with the inelastic scattering.

Lecture 24

Pion Form Factor

Data exists also for the pion form factor, for $q^2 < 0$ (spacelike)

$$\langle\pi^+|J_\mu|\pi^+\rangle = (p_1 + p_2)_\mu\ F_\pi(q^2) \qquad F_\pi(0) = 1 \tag{24.1}$$

It is best obtained from the reaction of $\gamma p \rightarrow n\pi^+$ produced by a virtual photon (generated by electron scattering).

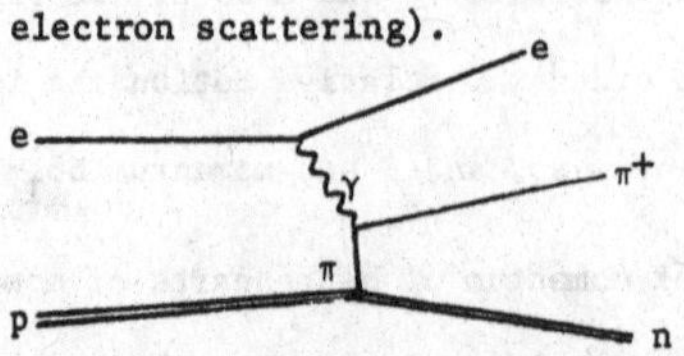

By going to sufficient energy one can obtain a region where virtual pion exchange is most important - and that this virtual π has as nearly zero q^2 (and is therefore as close to its mass shell as possible). The coupling of the virtual π to the nucleon is known in this region ($g_{N\pi\pi}\gamma_5$, $g^2 = 15$) so the only unknown is the photon-pion coupling between a real π and one that is nearly real, but a large momentum transfer Q.

Data is available out to $q^2 = -1.2$ where F_π has fallen to 0.28. This is just about exactly (for no known very good reason) the isovector part of the nucleon form factor $F_{1p}-F_{1n}$ which also fits all lower Q^2. It will also fit fairly well to the pure VDM or ρ meson pole $(1 + \frac{Q^2}{.56})^{-1}$, which is .32 at $Q^2 = -1.2$, but not at all well to the form factor G_D of q which is too low (.15 at $Q^2 = -1.2$).

We have already discussed this form factor for positive q^2 where it has been measured in the reaction $e^+e^- \to \pi^+\pi^-$. We have an experimental curve which is dominated by the ρ pole expression (with small understood corrections for the ω-ρ interference, etc.). It is interesting that an extrapolation so far into negative Q^2 still works reasonably well. The rate $e^+e^- \to \pi^+\pi^-$ measures, of course, only the absolute square $|F_\pi(q^2)|^2$ not the real and imaginary part separately. However the shape of the curve (ρ resonance) is so simple that our physical understanding may make a separation into real and imaginary parts not hopeless (e.g. we wrote previously, for the part near the resonance $(-m_\rho^2/(q^2-m_\rho^2 + i\Gamma_\rho m_\rho))$ and had some guesses as to how Γ varied with q^2. If this can indeed be done we would have information now known or soon available on $F_\pi(q^2)$ over a large range of q^2 both positive and negative. We can then check the hypothesis that everyone believes - that $F_\pi(q^2)$ must be an analytic function of its variable q^2 with a branch cut along the positive real q^2 - i.e. that F_π satisfies a dispersion relation.

$$F_\pi(q^2) = -\frac{1}{\pi}\int_{(2\mu)^2}^{\infty} \frac{\text{Im}\,F_\pi(q'^2)\,dq'^2}{q^2-(q')^2 + i\varepsilon} \qquad (\mu=m_\pi) \tag{24.2}$$

$F(q^2)$ has an imaginary part only for $q^2 > (2\mu)^2$ where at least $|F_\pi(q^2)|^2$ is measured. A check on a guess for the imaginary part would be to compute the real part from (24.2) for $q^2 > 0$ and check $|F_\pi|^2$ (alternatively a much easier equivalent method is to fit $|F_\pi|^2$ by choosing physically appropriate analytic expressions for the complex function $F_\pi(q^2)$). Then, having established a reasonable $\text{Im}\,F_\pi(q^2)$ one calculates $F_\pi(q^2)$ for negative q^2 (where it is real) to compare to experiment. We already see that it will work well - in the positive q^2 region the main feature is the ρ resonance and up to rather large q^2 no other resonance, in fact not a large production, has been found.

Below the inelastic threshold $q^2 = (4\mu)^2$ (and in practice probably some distance above it also) the phase of $F_\pi(q^2)$ must, by first order unitarity, be the same as the π-π scattering phase shift. Information on this is recently becoming available from pion-virtual pion scattering analyses. This may help in analyzing the dispersion relation.

How should we expect $F_\pi(q^2)$ to go for large q^2? It must fall off because the probability of producing just two π's where so many other states are available must be very low. The isospin current generated by the pair of pions must be coupled to other hadrons and we require this coupling not to radiate. As we shall understand better later, this leads one to expect F_π to fall as some power of q^2 for large q^2, but we do not know the power. However, these considerations make it not so suprising (granting the experimental behavior of $F_\pi(q^2)$ for positive q^2) that for negative q^2 even out to -1.2 the expression is close to the ρ pole formula (which arises from (24.2) if most of the imaginary part is near m_ρ^2).

Sum rules result from the following considerations. If the unsubtracted dispersion relation (24.2) is valid, since we know $F_\pi(0) = 1$ we must have

$$\frac{1}{\pi}\int_{(2\mu)^2}^{\infty} \frac{\text{Im } F_\pi(q'^2)\, dq'^2}{q'^2} = 1 \tag{24.3}$$

a "sum rule".

If we put $q^2 \to \infty$ in (24.2), we would expect $F_\pi(q^2)$ to go as $1/q^2$ times $-\frac{1}{\pi}\int \text{Im } F(q'^2)\, dq'^2$ (if $F_\pi(q^2)$ went more slowly than $1/q^2$ we would have to say $\int \text{Im } F(q'^2) dq'^2$ diverges, as indeed we would expect for Im $F(q'^2)$ would probably also fall slower than $1/q'^2$). If we expect $F_\pi(q^2)$ to fall faster than $1/q^2$ (we know no sure argument for this) we would have another relation

$$\int_{(2\mu)^2}^{\infty} \text{Im } F_\pi(q'^2) dq'^2 = 0$$

called a superconvergence relation.

We expect to have an unsubtracted dispersion relation valid because if (24.2) had an unknown constant added to the right-hand side then $F_\pi(q^2)$ could not approach zero as $q^2 \to \infty$. In any event we have the subtracted relation (24.2) - (24.3):

$$1 - F_\pi(q^2) = \frac{q^2}{\pi} \int_{(2\mu)^2}^{\infty} \frac{\text{Im}F_\pi(q'^2)dq'^2}{q'^2\,(q^2-q'^2)} \tag{24.4}$$

In practice (24.4) is more useful than (24.2) to compare $F_\pi(q^2)$ for negative q^2 to its experimental value since uncertainties from the contributions for large q'^2 where data is unavailable are much smaller.

We can now see using (24.4) why from what we already know experimentally for $F_\pi(q^2)$ for positive q^2 the value for negative q^2 should be close to the ρ pole value even rather far out. In (24.4) we know $\text{Im}\,F(q^2)$ is dominated by the ρ pole - there is not much $|F_\pi(q^2)|^2$ for larger q^2 (at least no resonance to 4GeV^2) and (24.4) does not contribute integrand for still larger q^2, for q^2 even as negative as -1.2. Thus $F_\pi(0) - F_\pi(q^2)$ is given by the analytic extension of the ρ pole formula nearly. But $F_\pi(0)$ is (the accident of the VDM constraints $f_{\rho\pi\pi}/g_\rho = 1$, see lecture 16) given by the ρ pole formula nearly, as we have seen earlier; thus $F_\pi(q^2)$ should also be. An interesting study would be to try to make this discussion quantitative. One can also ask if the data on $(e^+e^- \to \pi^+\pi^-)$ is accurate enough to also dominate (24.3) and thus explain the accident that $f_{\rho\pi\pi}/g_\rho = 1$.

<u>Proton Form Factor for Positive q^2</u>

The proton form factor for positive q^2 could be obtained from the reaction $e^+e^- \to p\bar{p}$. There are not yet any quantitative experiments but they are expected momentarily. One experiment near threshold saw no pairs, indicating G_M, G_E are small there - about as small (or smaller) than the dipole formula calculated there $q^2 \sim 4$. We would expect again that $F_1(q^2)$ and $F_2(q^2)$ are analytic functions satisfying a dispersion relation like

$$F(q^2) = -\frac{1}{\pi} \int \frac{\text{Im}F(q'^2)dq'^2}{q^2-q'^2+i\epsilon} \tag{24.5}$$

(When amplitudes can be expressed in more than one way, such as helicity flip amplitudes, or coeficients of Dirac matrices in spinor expressions - it is for the latter that the simplest dispersion relations are supposed to hold. Relations for other combinations must be deduced from these.)

This time we have a difficulty. $F(q^2)$ is measurable experimentally for $q^2 < 0$ by electron scattering and for $q^2 > 4M^2$ by e^+e^- annihilation into proton-antiproton pair. But how can it be defined experimentally in the region $q^2 = 0$ to $4M^2$? I do not know of a physical definition. But it is expected (fron field theory examples etc.) that $\mathrm{Im}F(q^2)$ is not zero in this region - in fact that the integral in (24.5) has a threshold at $(2\mu)^2$. Thus no amount of experimental data will permit a very detailed test of (24.5) for a piece of the integrand is completely unavailable. Of course, if a theory is available for $F(q^2)$ it can be defined in no-man's land $(0 < q^2 < 4M^2)$ by analytical continuation - but if we had a theory it would presumably already satisfy (24.5) and the proper thing to do would be to compare it directly to experiment in the physically available region.

Thus the dispersion relations for nucleon form factor are not of much direct use for comparing to experiment.

<u>NOTE</u>: We append here a theoretical note on the expected position of the threshold (24.5) below which $\mathrm{Im}\,F(q^2)$ can be expected to be zero.

For a charge density e^{-ar} classical physics gives $F = 1/(Q^2 + a^2)$ which has a first singularity at $Q^2 = -a^2$. A superposition of exponentials $\int_{\lambda_o}^{\infty} f(\lambda)e^{-\lambda r}d\lambda$ gives $\int_{\lambda_0}^{\infty} \frac{f(\lambda)}{Q^2+\lambda^2}\,d\lambda$ giving a continuum of singularities the lowest however being $Q^2 = -\lambda_0^2$, the slowest exponential tail. This idea is valid relativistically. (It has been shown that the position of the singularity is the same as the position of the singularities in Feynman diagrams of a process as if real particles were virtual).

For the proton the virtual state $N + \pi^+$ exists with a π extending out as (a gradient of) $e^{-\mu r}/r$. Its square, the charge density involves polynomials in $1/r$ times $e^{-2\mu r}$, the longest exponential is therefore 2μ, and the threshold at 2μ.

A more interesting case is one say like the Σ, of mass M_1 which virtually can emit a K (mass m) to be a nucleon mass M_2, with $M_2 < M_1$. We obtain the singularity by studying the perturbation diagram but can see its position physically this way:

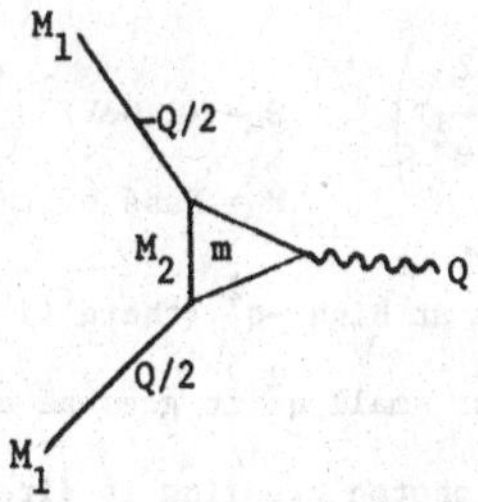

We ask: how sharp a rising (time independent) exponential electric potential can we tolerate before we get a divergence because the field rises faster than the charge density falls?

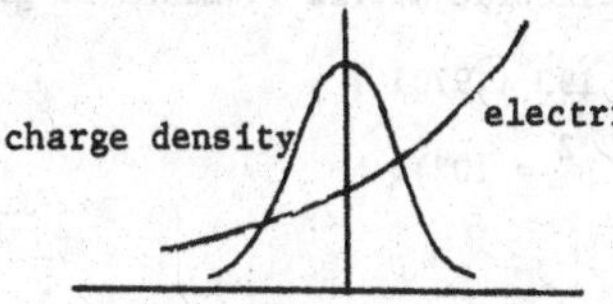

Thus we have a field of momentum Q (it is pure imaginary = ia) and are just able to make the K meson from the Σ. The virtual state is a particle of mass M_2 momentum 0, and one of mass m momentum + Q_2 where the initial state is one of mass M_1 of momentum G/q (Breit frame, ω = 0).

The condition of just divergence is that this state just has zero energy denominator

$$\sqrt{M_1^2 + Q^2/4} = \sqrt{M^2 + Q^2/4} + M_2$$

(this equation is relativistic because it conserves both energy and momentum). This has the solution

$$Q^2 = -a^2 = -4m^2 + (M_1^2 - M_2^2 - m^2)^2/M_2^2$$

if $M_2^2 < M_1^2 - m^2$ (otherwise the square roots are not positive.) The threshold is at $q^2 = +a^2$. That it is below the energy $4m^2$ needed to make a pair, which was the first guess, we call this an "anomalous" threshold.

For $M_2^2 > M_1^2 - m^2$ the equation cannot be solved, the singularity now really comes (normal threshold) when the photon can first make a pair $q^2 = 4m^2$.

Note on the quark model of Feynman, Kislinger and Ravndal.

The form factor for the proton by FKR is

$$\frac{G_M}{\mu} = (1 - q^2/4M^2)^{-1} \exp\left(\frac{q^2}{2\Omega}\left(1 - \frac{q^2}{4M^2}\right)\right) \qquad \Omega = (1\ \mathrm{GeV})^2$$

M = Mass of the proton

which is very poor - it cuts off too fast at high $-q^2$ (there is no reason to expect the model valid for large Q^2). For small q^2 it goes as about $1 - .7\,q^2$ instead of $1 + 2.6\,q^2$. In the model the photon coupling is direct (processes like VDM are not considered if they were we would add a factor $(1-q^2/m_\rho^2)^{-1}$ or $1 - 1.7\,q^2$ giving better understanding). The ratio $G_E\,\mu/G_M$ comes out $1 + q^2/2M^2$ or $1 + .5\,q^2$ which falls too fast with q^2 (experiment is around $1 + .06\ q^2$).

(Another way of dealing with the quark equations giving a much better form factor, but much worse fits to photon electric matrix elements is given in Fujimara et al. Prog. Theor. Phys. 44 193 (1970).)

For the pion FKR would get ($m_\pi^2 = \mu^2 = .02$).

$$F_\pi = \frac{1 + \frac{q^2}{4\mu^2}}{1 - \frac{q^2}{4\mu^2}} \exp\frac{q^2}{2\Omega}\left(1 - \frac{q^2}{4\mu^2}\right)$$

which is patently absurd (for example, having a zero at $q^2 = -0.08$!) It is by all odds the worst disaster of their model -(it comes from the ad hoc way they dealt with the spin of the quark unaffected by the photon) that one would think in honesty they would include it in their paper. They didn't because, surprisingly, they didn't think to calculate it when they wrote their paper!

Electron-Proton Scattering. Deep Inelastic Region

Lecture 25

Other Photon Processes for $q^2 < 0$

Electron beams permit us to study reactions such as $e\ p \to e\ x$ where x is some hadron final state.

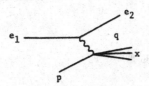

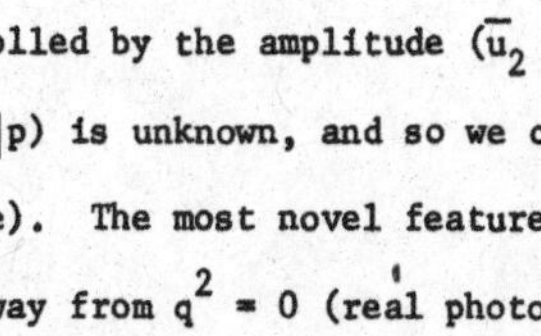

It is controlled by the amplitude $(\bar{u}_2\ \gamma_\mu\ u_1)\ \frac{4\pi e^2}{q^2}\ (x|J_\mu|p)$ in which only the factor $(x|J_\mu|p)$ is unknown, and so we can measure this matrix element (or rather its square). The most novel feature is the ability to vary the q^2 of the virtual photon away from $q^2 = 0$ (real photon experiments, which we already discussed) to q^2 negative, and in fact to the far negative region. In addition, very energetic (high mass) states x can be excited in the high energy machines. A great deal of information is becoming available, and there are many theoretical questions especially for the new large q^2 and high mass states x.

The simplest experiments to do are those which do not require studying the x states at all (sum over possible states x) but just study the electron beam -

its deflection and energy determines the momentum transfer and its energy loss determines (in the lab system) the energy of x above the proton mass. These experiments were done first and are most complete - we discuss first therefore their results and theoretical interpretation (which will occupy us for many lectures). Then we shall return and discuss what we can expect the final states x to look like and how they do behave experimentally as far as is known.

Inelastic Electron Nucleon Scattering

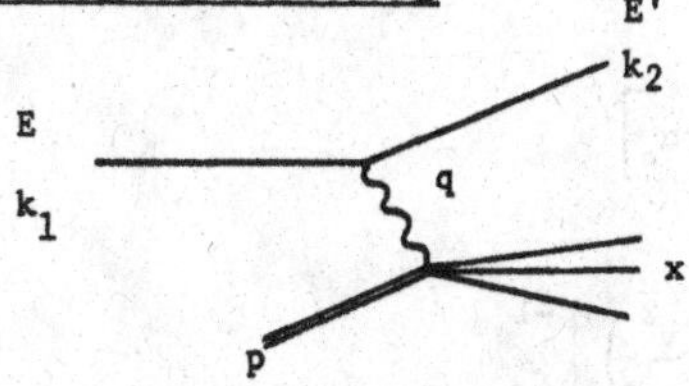

The energy loss $\nu = E-E'$ in the lab. of the electron is measured as well as its 4-momentum transfer q^2. The invariants of the problem are thus q^2 and $p \cdot q = M\nu$ where p is the 4-momentum of the proton. All quantities measured proper to the hadrons are expressed as functions of these two variables. The probability we measure is then proportional to

$$M^2 = (\bar{u}_2 \gamma_\nu u_1)^*(\bar{u}_2 \gamma_\mu u_1)\left(\frac{4\pi e^2}{q^2}\right)\sum_x \langle p|J_\nu(-q)|x\rangle\langle x|J_\mu(q)|p\rangle \; 2\pi\delta((p+q)^2 - M_x^2) \qquad (25.1)$$

In the case of unpolarized electrons (there are no experiments with polarized electrons and protons yet, but we shall discuss the theory later), the sum and average over electron states contributes the spur $(\bar{u}_2 \gamma_\nu u_1)(\bar{u}_2 \gamma_\mu u_1) =$ $1/2 \; sp\,(\not{k}_2 \gamma_\nu \not{k}_1 \gamma_\mu) = 2\,(k_{1\mu}k_{2\nu} + k_{1\nu}k_{2\mu} - \delta_{\mu\nu}\, k_1 \cdot k_2)$. While the hadron state sum gives an expression we shall call $K_{\mu\nu}$

$$K_{\mu\nu} = \sum_x \langle p|J_\nu(-q)|x\rangle\langle x|J_\mu(q)|p\rangle \; 2\pi\delta\,(M_x^2 - (p+q)^2) \; . \qquad (25.2)$$

(spin of p averaged, is also $\mu\nu$ symmetric part of expression for any spin). (Note, the mass^2 of the final states which we are generating is expressed in terms of our variables q^2, ν as

$$M_x^2 = (p+q)^2 = M^2 + 2\,M\nu + q^2 \qquad q^2 \text{ is negative.}$$

By varying ν for given q^2 we can vary M_x^2 and look, for example, for final state resonances.)

From arguments of gauge invariance and relativity the symmetrical part of the tensor $K_{\mu\nu}$ (which is all that counts for unpolarized electron, and is all that survives if the proton is unpolarized) must have the form

$$\frac{M}{\pi} K_{\mu\nu} = 4\, W_2(q^2,\nu)\, \left(p_\mu - q_\mu \frac{p\cdot q}{q^2}\right) \left(p_\nu - q_\nu \frac{p\cdot q}{q^2}\right) - 4\, W_1(q^2,\nu)\, M^2 \left(\delta_{\mu\nu} - \frac{q_\mu q_\nu}{q^2}\right) \tag{25.3}$$

Several properties of W_1, W_2:

The apparent, but unwanted poles at $q^2 = 0$ come from the way we have written our expression and of course cannot be real. To remove the highest pole we must have $(p\cdot q)^2 W_2 = -M^2 q^2 W_1$ as $q^2 \to 0$ or $W_2 = -q^2/\nu^2\, W_1$ + order q^4 as $q^2 \to 0$, but W_1 can approach a constant $W_1(0,\nu)$ as $q^2 \to 0$. Hence for small q^2, $W_1 = W_1(0,q^2)$; $W_2(q^2,\nu) \to -q^2/\nu^2\, W_1(0,\nu)$.

The direct total absorption cross section for an imaginary photon of mass q^2 to hit a proton would be

$$\sigma = \frac{4\pi e^2}{2k2M} \sum |\langle x|e_\mu J_\mu|p\rangle|^2\, 2\pi\delta\left(M_x^2 - (p+q)^2\right)$$

$$= \frac{4\pi e^2}{2k2M}\, e_\mu e_\nu K_{\mu\nu} \tag{25.4}$$

k = momentum of photon in p rest system.

We have two polarization cases:

Transverse. e_μ is perpendicular to q_μ and to p_μ

$$\sigma_t = \frac{4\pi^2 e^2}{k}\, W_1 \tag{25.5}$$

Longitudinal. Sometimes called scalar

$$e_t = q_z/\sqrt{-q^2}, \quad e_z = q_t/\sqrt{-q^2}$$

$$\sigma_s = \frac{4\pi^2 e^2}{k}\left[\left(1 + \frac{\nu}{-q^2}\right) W_2 - W_1\right] \tag{25.6}$$

These two cross sections can be used as parameters instead of W_1, W_2. From the above formulas

$$\frac{W_1}{W_2} = \left(1 + \frac{\nu^2}{-q^2}\right)\left(\frac{\sigma_t}{\sigma_t + \sigma_s}\right) \tag{25.7}$$

It is convenient to define

$$R = \sigma_s/\sigma_t\,.$$

Combining the electron spur to the expression for $K_{\mu\nu}$ we obtain the total cross section in the lab.

$$\frac{d^2\sigma}{d\Omega dE'} = \frac{\alpha^2}{4E^2 \sin^4 \theta/2} \left[(\cos^2 \theta/2)\, W_2 + 2\, (\sin^2 \theta/2)\, W_1\right] \tag{25.8}$$

In principle it is possible to get both W_1, W_2, this is done in several cases - but most accurate is W_2 and often what is given is W_2 under some assumption about W_1. For example we have

$$\frac{d\sigma}{d\Omega dE'} = \frac{e^2}{2\pi^2} \frac{k\, E'}{-q^2 E(1-\varepsilon)} \left[\sigma_t + \varepsilon\sigma_s\right] \tag{25.9}$$

$$\varepsilon = \left[1 + 2\tan^2 \theta/2 \left(1 + \frac{\nu^2}{-q^2}\right)\right]^{-1}$$

and often data is given for $\sigma_t + \varepsilon\sigma_s$ without resolving them, ε is often small.

Problem: Show that for proton elastic scattering

$$4M^2 W_1 = -q^2\, G_M^{\;2}\, \delta(\nu + q^2/2M)$$

$$W_2 = \frac{G_E^{\;2} - \frac{q^2}{4M^2} G_M^{\;2}}{1 - \frac{q^2}{4M^2}}\, \delta\,(\nu + q^2/2M)$$

Lecture 26

Inelastic Electron Nucleon Scattering (continued)

We shall now give a kind of preliminary description of the experimental results - some details will be dealt with further later. First a useful variable to discuss the lower energy results in is the mass^2 of the final state $M_x^{\;2}$. $M_x^{\;2} = M^2 + 2M\nu + q^2$. We shall, at first, use $M_x^{\;2}$ and q^2 in our discussion. As a function of $M_x^{\;2}$ we, of course, get nothing below M^2 then a large elastic peak corresponding to $M_x^{\;2} = M^2$ when the final state is just a proton - proportional to $\delta(q^2 + 2M\nu)$ - and depending on G_M, G_E as we have previously studied. This contribution falls as q^2 rises as $1/(1 - q^2/.71)^2$.

Next, for fixed q^2, as $M_x^{\;2}$ varies, we see a "first" resonance at $M_x = 1236$, a "second" one near M_x = 1520 and finally a "third" around 1700 and some indication of a fourth at 1900. Theoretically we should see more resonances, e.g. 1535 but undoubtedly the "second" peak is an unresolved mixture of these two while the one near 1700 sometimes called the 1688 resonance is probably that with four others expected near that energy. The resonance at 1407 (the Roper resonance) has not been seen in these experiments.

These resonances can be fitted as Breit-Wigner peaks on a background which gradually dominates into a smooth curve as M_x^2 increases.

How do the resonance strengths vary with q^2? At very low q^2 the behavior depends theoretically on the angular momentum of the state and starts as an appropriate power of Q ($Q^2 = -q^2$). For higher q^2, however, the strength of the resonances all fall more or less as does the elastic peak. In fact if one plots the ratio $(d\sigma/d\Omega)_{res}/(d\sigma/d\Omega)_{elastic}$ with $-q^2$ one finds curves which rise rapidly from threshold (photoelectric) values to saturation (in the vicinity of one or just below) for q^2 > about 1 GeV^2.

For larger q^2 and large ν there is nothing left of the resonances. There it was suggested by Bjorken that the function $\nu W_2(q^2,\nu)$ (and also $W_1(q^2,\nu)$) should be a function of the variable $x = -q^2/2M\nu$ only. (Data is often presented also as a function of $\omega = 2M\nu/-q^2 = 1/x$). This has turned out to be rather accurately true. This feature, that as $-q^2 \to \infty$, $\nu \to \infty$ such that $-q^2/2M\nu = x$ is kept fixed, $\nu W_2(q^2,\nu)$ approaches a function of x, F(x) is known as Bjorken scaling. We shall have a great deal to say later of its theoretical significance.

Present (January 1972) data is available in the following regions.

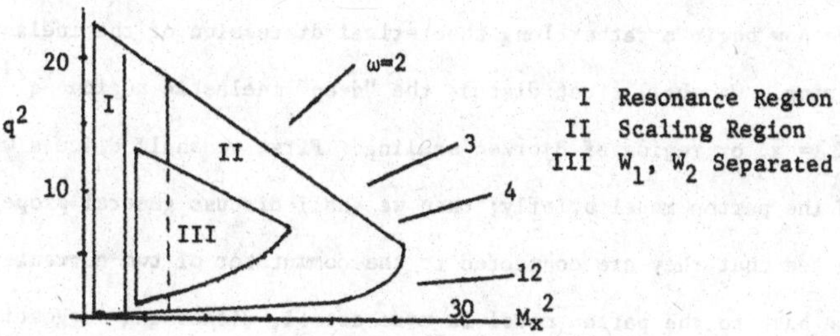

In the region $\omega < 4$, scaling is observed above $M_x^2 = (2.6)^2$. In the region 12 > ω > 4 for $M_x > 2$ $-q^2 > 1$; νW_2 is a constant within errors and scales in any variable.

The data for νW_2 looks as follows

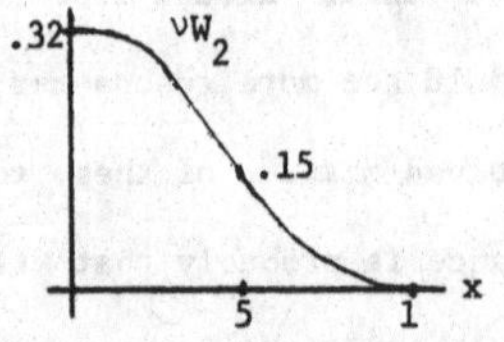

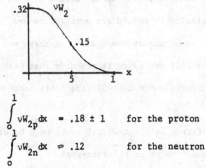

$$\int_0^1 \nu W_{2p} dx = .18 \pm 1 \qquad \text{for the proton}$$

$$\int_0^1 \nu W_{2n} dx \approx .12 \qquad \text{for the neutron}$$

Data for the neutron (by subtraction from the deuteron) is also available. The ratio to W_{2p} looks like

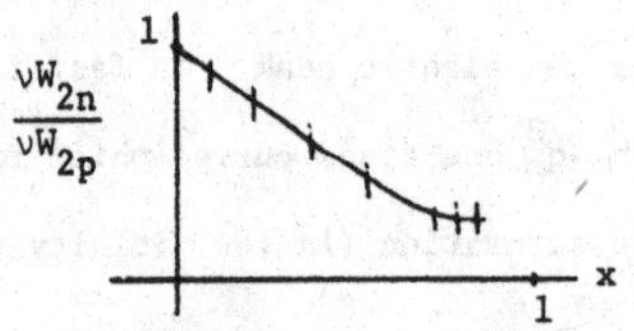

It is consistent with a straight line $(1 - x)$ from $x = 0$ to $.75$. There are indications that it curves away above the line for larger x and could reach a value as high as .4(?) for $x \to 1$, but higher values are probably ruled out.

In the region where W_1 and W_2 can be separated, values of R can be found - they vary from 0 to 0.5 with large errors so no definite trend can be seen. If R is assumed constant it is $.18 \pm .10$, the data is also compatible with $R = .03\ (-q^2)/M^2$ or with $R \approx q^2/\nu^2$. In the latter case we would expect with scaling that $R \to 0$ for fixed x.

Theory of the Inelastic Electron Proton Scattering

We now begin a rather long theoretical discussion of the inelastic e - p scattering. We shall first discuss the "deep" inelastic region $-q^2 \to \infty$, $\nu \to \infty$, $-q^2/2M = x$, or region of Bjorken scaling. First we shall discuss what is called the parton model briefly; then we shall discuss general properties of W_2, W_1 and see that they are connected to the commutator of two currents. Finally we come back to the parton model in more detail, discussing suggestions that the partons be identified with quarks, etc. - and then discuss the relation of this to a more abstract representation called Gell-Mann's light cone algebra. There then follows miscellaneous discussion, ending with attempts to understand things in other regions of the q^2,ν plane (i.e. for smaller q^2, and, or at resonances) etc., (actually our discussions will not be so neatly organized, we shall wander among these subjects).

Parton Model

Parton Model

The parton model is the conceptually easiest to understand, although it appears a bit special, as special assumptions seem to be made. More general abstract considerations are therefore more satisfying, but first we discuss the more elementary viewpoint. In discussing these matters it is best to keep in mind all the principles of relativity, quantum mechanics, unitarity, analyticity, etc. One way to do this is to work in a conceptual model which satisfies all these principles simultaneously. There is no known simple model which does this except field theory (and that may not do it - all examples diverge!), and that is a very complicated model indeed. Nevertheless we shall try to see what field theory might suggest.

In field theory the wave function for a state, such as a proton, could be given by giving the amplitudes to find various kinds of bare field particles moving with various momenta. These bare field particles we call "partons". Sometimes some phenomena can be understood directly in terms of this wave function but usually matrix elements to wave functions of other states must be analyzed. The wave function, however, is not easily transformed from one relativistic system to another (the Hamiltonian must be known) because it represents a slice at a given time. Certain properties are therefore more easily seen from the wave function in one system than in another. The deep inelastic scattering behavior is best understood from the wave function for the proton with extremely high momentum P (in +z direction). In fact we study the limiting form of the wave function as $P \to \infty$.

In what variables will the wave function have a limiting form? From a study of the character of typical field hamiltonian in a few examples, and from a study of very high energy inclusive hadron collisions we conclude that if the momentum of the parton is measured proportional to P as ξP and the transverse momentum in absolute units k then the amplitudes depend only on ξ, k as P rises - (except for ξ so small that ξP is a few GeV, ξ of order 1 GeV/P are called "wee" - the distribution of "wee" partons is probably best described in absolute p_z k variables. As we shall see the "wee" region is hard to analyze - but the main features of the deep inelastic scattering does not involve them).

The fact that $k_\perp$ is finite is not, at least in any obvious way, a direct consequence of field theory - (in fact perturbation theory does not give this result, and therefore must not be reliable here) it is simply guessed at from the ubiquitous result of high energy collissions that the transverse momenta available to the products averages about $\overline{p_\perp^2} = (.4\ \text{GeV})^2$.

But granting this, the ξP scaling is expected from field theory (although today we could base this also directly on experiment, it is the scaling law for the longitudinal momentum of the products in very high energy collisions). Some suggestion of how it works is this. The amplitude that a state of energy E is seen to be made up of two parts of energy $E_n = E_1 + E_2$ is dominated (in perturbation theory) by a factor

$$A/(E - E_n) = A/(E - E_1 - E_2) \tag{26.1}$$

But the z component of momentum of the parts is that of the whole $P = p_1 + p_2$. Set $p_1 = \xi_1 P$, $p_2 = \xi_2 P$, $\xi_1 + \xi_2 = 1$, $E = \sqrt{P^2+M^2} \approx P + \frac{M^2}{2P}$. If M_1 is the mass of the first part and k_1 its transverse momentum, we have

$$E_1 = \sqrt{\xi_1^2 P^2 + k_1^2 + M_1^2} \approx \xi_1 P + \frac{k_1^2+M_1^2}{2\xi_1 P} \quad \text{approximately.} \tag{26.2}$$

Likewise

$$E_2 \approx \xi_2 P + \frac{k_2^2+M_2^2}{2\xi_2 P}$$

$$\text{Therefore } A/(E - E_1 - E_2) = AP/\left(\frac{M^2}{2} - \frac{M_1^2+k_1^2}{2\xi_1} - \frac{M_2^2+k_2^2}{2\xi_2}\right) \tag{26.3}$$

(The P in the numerator is usually absorbed in normalization or in the form for A, the essentially complicated behavior of this is a function of ξ_1, ξ_2, k_1^2, k_2^2. Arguments like these were originally used to predict the inclusive scaling, and they have been confirmed by experiment.

Furthermore we see ξ_1 (outside the wee region) must be positive - no partons are going vigorously backward - because (26.2) is not valid if ξ_1 is negative, it should be $|\xi_1|$ on the right hand side. Then (if ξ_1 is negative) $E - E_1 - E_2 \approx (1-|\xi_1|-\xi_2)P = 2\xi_2 P$ so the denominator is not small but large and the amplitude is $1/P^2$ smaller than the preceeding case.

Lecture 27

Parton Model (continued)

We envisage the proton of momentum P as being made of partons of momenta $\xi_i P$ all sharing in various proportions ξ_i the momentum of the proton, all ξ lie between 0 and 1 (else some other would have to be negative since $\sum \xi_i = 1$).

We shall therefore think of the incoming proton as a box of partons sharing the momentum and practically free. Another way to look at this is to take a dynamic view of the parts in the rest system and assume finite energy of interaction among parts, so as time goes on they change their momenta, are created or annihilated, etc., in finite times. But moving at large momentum P these times are dilated by the relativistic transformation so as P rises things change more and more slowly, until ultimately they appear not to be interacting at all.

When the proton is hit by the photon the interaction operator J_μ couples to one parton or another and knocks it to a new state of momentum $p_2 = p_1 + q$ ($p_1 = \xi P$ nearly). If the parton had mass m^2 we would expect a rate proportional to

$$\frac{1}{2E_1}|M|^2\ 2\pi\delta((p_1+q)^2 - m^2) = \frac{1}{2E}\ K_{\mu\nu} \tag{27.1}$$

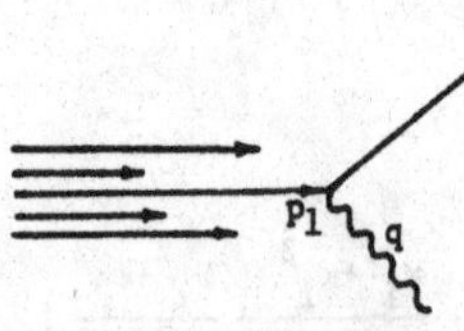

(the factors $1/2E_1$ and $1/2E$ are included because of the normalizations of $|M|^2$ and $K_{\mu\nu}$). In fact, due to the interaction energies assumed finite, the energy loss is not just that for a free particle but differs from it by an unknown (but finite which may vary with x, etc.)

For spin zero partons

$$|M|^2 = (p_{1\mu} + p_{2\mu})\,(p_{1\nu} + p_{2\nu}). \tag{27.2}$$

For spin 1/2 partons

$$|M|^2 = 1/2\ \mathrm{sp}((\not p_1 + m)\,\gamma_\mu\,(\not p_2 + m)\gamma_\nu)$$

$$= (2\,p_{1\mu}\,p_{2\nu} + 2\,p_{1\nu}\,p_{2\mu} - 2\,\delta_{\mu\nu}(p_1\cdot q)) \tag{27.3}$$

(We shall calculate for spin 1/2, just state results for spin 0.) To make things easy we will omit all terms in q_μ in $K_{\mu\nu}$ because we know how to get them by gauge invariance. Thus in $|M|^2$ put $p_{1\nu}$ in place of $p_{2\nu}$. In our limiting circumstances we can write $p_{1\mu} = \xi P_\mu$ to an excellent approximation where P_μ is the four-momentum of the proton (strictly it is valid for the z component only, but we have seen that implies it for the t component $E_1 = \xi E$; and the transverse components being only finite are relatively small).

Therefore if

f(x)dx = number of partons with momentum between x and x+dx each weighed by the charge squared (in units of electron charge) we have

$$K_{\mu\nu} = \frac{\pi}{M}(4\,P_\mu P_\nu W_2 - 4\,\delta_{\mu\nu}\,M^2 W_1)$$

$$= \int \frac{f(\xi)}{\xi}\,(4\,\xi^2\,P_\mu\,P_\nu - 2\xi\delta_{\mu\nu}P\cdot q)\;2\pi\delta((\xi P+q)^2 - m^2 - \Delta)\,d\xi \tag{27.4}$$

(the first $1/\xi$ comes from the normalizing factors $2E/2E_1$). Inside the δ function we have $(\xi P+q)^2 - m^2 - \Delta = 2\xi\,(P\cdot q) + q^2 + \xi^2 M^2 - m^2 - \Delta$; but as $-q^2 \to \infty$ and $P\cdot q = M\nu$ goes to ∞ with $-q^2 = 2\,M\nu x$ this becomes $2\,M\nu\,(\xi - x)$ + (finite) or nearly $\delta\,(2\,M\nu\,(\xi-x)) = \frac{1}{2M\nu}\,\delta\,(\xi-x)$. Therefore we have, integrating the δ function, for large ν

$$4\,P_\mu P_\nu W_2 - 4\,\delta_{\mu\nu}\,M^2 W_1 =$$

$$= \frac{f(x)}{\nu}\,(4\,x\,P_\mu P_\nu - 2\,M\nu\,\delta_{\mu\nu}) .$$

Thus

$$\nu W_2(q_1^2\,\nu) = x f(x)$$

$$2MW_1(q_1^2\ \nu) = f(x) \tag{27.5}$$

and we have functions of $x = -q^2/2M\nu$ only.

If we used scalar partons, the formula for νW_2 is unchanged but $W_1 = 0$. If the fraction of the partons which a spin 1/2 (weighed by their charge squared) at momentum x is $\gamma(x)$ we get W_2 unchanged, $2MW_1 = \gamma(x)\ f(x)$. The ratio

$$R = \frac{\sigma_s}{\sigma_t} = \frac{\left(1+\frac{\nu^2}{-q^2}\right)W_2 - W_1}{W_1}$$

in the scaling limit would be

$$R = \frac{\nu W_2/x - 2MW_1}{2MW_1} = \frac{1-\gamma(x)}{\gamma(x)} \tag{27.6}$$

In the region studied R is of order .18 ± 10 so γ is less than .2, not many scalar partons. A much more likely hypothesis is

<u>The current carrying partons are all spin 1/2</u>

The value of R remaining today experimentally being due to our having not enough energy and q^2 to be fully in the scaling limit. E.g., $R \approx q^2/\nu^2$ would also fit the data and give zero in the scaling limit.

This is a very profound conclusion about the structure of the underlying theory for hadrons. We must watch to see if R really does approach zero as the scaling limit is reached.

If charged partons all carried the fundamental charge ±e then charge squared is 1 in our units and thus we could say only 18% of the momentum is carried by charged partons in the proton (because $\int xf(x) = .18 \pm .01$) - the remainder, 82% would be carried by neutrals. This 18% is surprisingly small. If the partons are quarks and carry charges like ±2/3 or ± 1/3 the percentage of momentum carried by the quarks could be higher. We shall discuss such a model in more detail, but it turns out even then it is necessary to assume something else neutral carries part of the momentum.

Lecture 28

<u>The Wee Region</u>

The ideas leading to the scaling formulas (27.5) are very reliable. By making further assumptions about the wee region we can understand other aspects of W_1 and W_2. It must be realized we are now elaborating on our original parton ideas developing them further to understand more features of the f(x) curve. The data for νW_2 seems to approach a constant (.32) as $x \to 0$ which means that f(x) would go as .32/x showing a mean number of partons rising as x falls into the wee region (such that the number of wee partons is finite and independent of P, and the mean total number of partons in a state of momentum P rises logarithmically with P). This is not entirely unexpected, it is the same as the distribution of products in hadronic collisions. The way this appears to happen can be gathered by studying the field theory equations at high energy, and also the perturbation theory of bremsstrahlung. In the latter case neutral particles are generated in a $dk/E = \sim dP_z/\sqrt{P_z^2+k_\perp^2+m^2} \to dx/x$ distribution. These neutrals can generate pairs so the small x region contains large and nearly equal numbers of particles and antiparticles. The field equation approach suggests the same thing; and further that the low region is generated in higher order perturbation from the higher momenta by a series of cascades $x \to x' \to x''$ going down in x. In either case we conclude the character of the pairs will be as a whole neutral and therefore the same for proton and neutron. These expectations have been made by other means also, for example, the dx/x leads to a constant cross section for virtual photons (of fixed but large, negative mass squared). That these cross sections should be a constant is expected from considerations that these photons have an amplitude to be virtual hadrons (like the ρ) and hadrons give constant cross sections (the magic word "Pomeron" is used to "explain" this), the same for proton and neutron. Thus we expect νW_{2n} to equal νW_{2p} as $x \to 0$ as indeed they turn out to do; both f(x) for p and n approach the same .32/x for small x. We also know experimentally that at $q^2 = 0$ $\sigma_{\gamma p} = \sigma_{\gamma n}$ to 3% at 16 GeV.

The fact that f(x) seems to go as .32/x as $x \to 0$ implies that the total cross section for virtual photons of energy ν on protons is a constant as energy goes to ∞ for fixed and large negative q^2, just as the real photon cross section does. We can expect, as long as we are on the high side of the wee region, that $x = b/P$ with b large enough that, as our measurements tell us, $f(x) \approx .32/x$. Hence the transverse virtual photon cross section is

$$\sigma_t = \frac{4\pi e^2}{k} W_1 = \frac{4\pi^2 e^2}{\sqrt{\nu^2 - q^2}} \frac{1}{2M} f(x) = \frac{4\pi^2 e^2}{\nu} \frac{f(x)}{2M} \tag{28.1}$$

in the scaling limit. And for x on the high side of the wee region ($-q^2$ large enough,)

$$\sigma_t = \frac{4\pi^2 e^2}{2M\nu}\ \frac{.32}{x} = \frac{4\pi^2 e^2(.32)}{-q^2}$$

$$= 115\ \mu b\ \left(\frac{.32\ \mathrm{GeV}^2}{-q^2}\right) \qquad (28.2)$$

At this point we may well guess (these considerations are independent of the parton model) what happens for ν large but $-q^2$ is not large enough that $-q^2/2M\nu$ is far enough out of the wee region ~ 1 GeV/P. We know for $q^2 = 0$ the total photo-cross section is independent of ν. For other $-q^2$, the unreliable VDM would give a factor $-m_\rho^2/(q^2-m_\rho^2)$ in the amplitude or $(1/(1-q^2/m_\rho^2))^2$ in the cross section times a ν independent term (large ν). This is clearly wrong for large $-q^2$ (far from the ρ pole) for it falls as $(-q^2)^{-2}$, instead of $(-q^2)^{-1}$ as we have just seen is experimentally found, above. But we can certainly guess that, for each $-q^2$ the cross section is constant but dependent on $-q^2$ for large ν, $\sigma_t = 4\pi^2 e^2\, C(-q^2)$. We do not at present know $C(-q^2)$ except for $q^2 = 0$ and large $-q^2$. We need a good theory for this function.

Although the number of partons is infinite, the momentum contained in them is finite, of course, because the total momentum of all partons neutral and charged is 1 (units of P). We have for the momentum carried by all charged partons weighed by $e^2 \int_0^1 xf(x)dx = .18 \pm .01$.

Formula for R

We may get an idea of how R behaves as we approach the scaling limit by calculating σ_s directly in this region. We work in the system in which q is pure space-like

$$P_\mu = (E, P, 00) \qquad P_\mu - \frac{P\cdot q}{q^2} q_\mu = (E, 0, 0, 0)$$

$$q = (0, -2Px, 0, 0) \qquad \delta_{\mu\nu} - \frac{q_\mu q_\nu}{q^2} = \mathrm{diag.}\ (1, 0, -1, -1)$$

Use Equation (27.4) with all q_μ put back into $|M|^2$

$$\frac{\pi}{M}\ (4P_\mu P_\nu W_2 - 4\delta_{\mu\nu} M^2 W_1) =$$

$$\int \frac{f(\xi)}{\xi}\ 2[p_{1\mu} p_{2\nu} + p_{1\nu} p_{2\mu} - \delta_{\mu\nu}(p_1\cdot p_2 - m^2)]\ 2\pi\delta((p_1+q)^2 - m^2 - \Delta)d\xi$$

(the expression in the δ function becomes $(4\,P^2\,x(\xi-x)-\Delta)$). Now to get σ_s we want the t component of both sides; however, if the RHS were exact the z component would exactly vanish (from gauge invariance). To avoid errors of differences of large quantities we take μ = t-z, ν = t-z so $\delta_{\mu\nu}$ vanishes, $p_{1\mu}\,p_{2\nu} \to (\varepsilon_1 - p_{1z})(\varepsilon_2 - p_{2z})$ (this is equivalent to replacing the polarization of the photon $e_\mu = (1, 0, 0, 0)$ by $e_\mu{}' = (1, 1, 0, 0)$; we should get the same result by gauge invariance). Now p_{2z} runs backwards so $\varepsilon_2 - p_{2z} \approx 2|p_{2z}| = 2(2x-\xi)p$. But p_{1z} runs forward so $\varepsilon_1 - p_{1z} \approx \sqrt{p_{1z}^2 + k_\perp^2 + m^2} - p_{1z} \approx \frac{k_\perp^2+m^2}{2\,p}$. Here $k_\perp^2+m^2$ is the mean perpendicular momentum plus the mass^2 of the parton (at ξ), whatever that means, but at least it is the square of some finite energy. σ_s and R are given by

$$(4E^2W_2 - 4\,m^2W_1) = 2Mf(x)\,\frac{4(k_\perp^2+m^2)}{-q^2}\Bigg|_{\text{at } x}$$

$$R = \frac{4(k_\perp^2+m^2)}{-q^2}\Bigg|_{\text{at } x'}$$

This explicit dependence on m^2 is erroneous. The deviations in the parton model of calculations using free partons are probably to set $\pm\Delta$ (a binding energy correction) uncertainties on all effective parton masses. In this special case the error comes in assuming in $\varepsilon_1 - p_{1z}$ that ε_1 is just the kinetic energy of the free parton and not more complicated. When operators involve d/dt in Schroedinger perturbation theory their exact expression is usually more complicated than just the kinetic energy operator. The formula for R should be

$$R = \frac{4(k^2+m^2\pm\Delta)}{-q^2}\Bigg|_{\text{at } x}, \tag{28.3}$$

The expected falling trend with $1/(-q^2)$ is not evident in the data, but the errors are large. If we estimate $k^2 \approx .25(\text{GeV})^2$ independent of x as in hadronic collisions, and m^2 of the same order (else we could not explain why $k_\perp^2$ is so small), we find $R \approx (2\pm4\Delta)/(-q^2)$; for $-q^2 = 7$ (near center of data) $R \approx .3 \pm 4\Delta/7$ not unreasonable compared to the average .18.

Lecture 29

The region near $x = 1$

We can get some qualitative ideas of the region near $x = 1$ by first studying the extreme case $x = 1$, the elastic form factors, at large $-q^2$. We take the coordinate system with $q = (0, -Q, 0, 0)$ a pure space-like vector. P_μ the momentum of the proton $= (E, P, 0, 0)$ must have $P = Q/2$, after the collision the proton has momentum $P'_\mu = P_\mu + q_\mu = (E, -P, 0, 0)$, with z component in the opposite direction. For large Q, hence large P, we can describe the initial state of the proton as an amplitude for various configurations (in x, $k_\perp$). Let us suppose for example, the configuration contains two non-wee partons and two wee, and draw a picture:

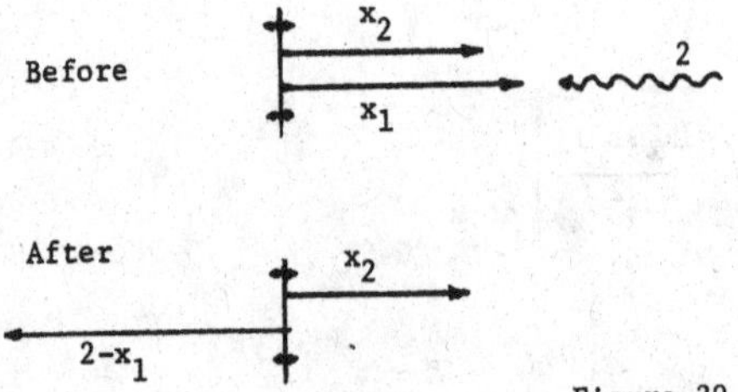

Figure 29.1

Now we ask that the final state (the "after" picture) be only a proton - that is, a factor <Proton at $-P$ | "after state">, the (complex conjugate of) amplitude that the after state appears in the proton wave function. But a left moving proton can never (for large P) look like the after picture because it contains a backward moving parton (x_2), (and any estimate of its mass square is order P^2, requiring order P^2 binding energy contributions to get the mass^2 down to M_p^2).

Thus insofar as the proton looks like "before" it contributes very little to elastic scattering. To scatter elastically we must start with a configuration where one parton has nearly all the momentum, $x = 1$ and all the rest are wee, say below $1/P$.

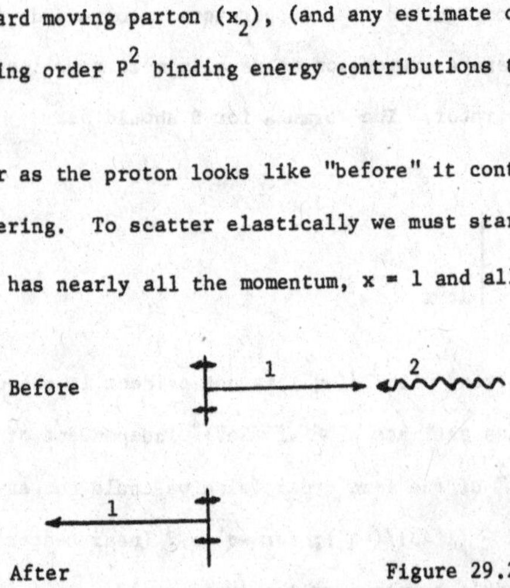

Figure 29.2

Now the single fast one is reversed by the incoming photon, the energy is not changed much, and the final configuration is possible for a proton because wee partons (those with absolute p_z finite say below order 1 GeV) can go backwards as well as forwards. Therefore the scattering amplitude (form factor) is proportional to the square of the amplitude that a proton looks like "before", hence proportional to the probability that the proton has all partons but one in the wee region.

We now see qualitatively why this should be some inverse power of Q in the following manner. If the bremsstrahlung analysis of the "sea" (dx/x) region has any qualitative reality we can judge the probability that no particles have their x above x_o of order 1/Q to be $e^{-c\bar{n}}$ where $\bar{n}$ is the mean number expected above this x_o and c is some constant to allow for correlations (unlike first order bremsstrahlung which is purely Poisson, perhaps each "photon" could make two of some other object, etc., so that the mean number of particles would be higher but proportional to some vaguely defined "mean number of statistically independent events"). Now we have seen, for small x at least, n goes as a/x where a is another constant - hence $\bar{n}$, the mean number above x_o is $a\int_{x_o}^{x_1} \frac{dx}{x} = a \ln x_1 - a \ln x_o$ where x_1 is some unknown upper limit where the dx/x ideas break down. Thus for small enough x_o, $e^{-c\bar{n}}$ varies as exp (ca (ln x_o - ln x_1)) as a constant times a power of x_o. That is, the probability no "sea" particles are outside the wee region falls as a power of Q.

There may be some special partons to make up total quantum numbers which are not parts of the sea which must also be in the wee region. But scaling studies of the field equations as well as other theoretical arguments (related to Regge-type behavior) suggest very strongly that they would show behavior of the form x^α dx/x where $\alpha \geq 0$, and thus their probability to be found in the region below x_o would also fall as a power of x_o.

We conclude that the probability the proton looks like our parton near x = 1 with all the others below some low limit x_o falls as a power of x_o, say as x_o^γ.

In particular then for x_o wee of order 1/Q, we find the probability varying as $Q^{-\gamma}$.

Thus the form factor G_M of a proton (and G_E also, see below) should fall as a power of Q, $Q^{-\gamma}$. Experimentally γ is probably a little larger than 4.

The fact that the form factor falls toward zero as $Q \to \infty$ means that a proton (as $P \to \infty$) has a zero amplitude to appear as a single parton. This may be because no single parton has the precise quantum numbers of the proton, but even if it did, a more cogent reason is that the probability of finding a parton without other fields generated by interaction from it (represented by the sea dx/x) is zero, (just as is expected from the theory of bremsstrahlung).

Our analysis of the ratio $R = \sigma_s/\sigma_t$ described in the previous lecture can be as well applied to the elastic piece $x = 1$. We therefore find

$$\left.\frac{\sigma_s}{\sigma_t}\right|_{\text{elastic}} = \left.\frac{4(k_\perp^2 + m^2 \pm \Delta)}{-q}\right|_{x = 1}$$

where $k_\perp^2 + m^2 \pm \Delta$ is the value for the single parton with nearly all the momentum. But this ratio expressed in terms of G_E, G_M is simply

$$\left.\frac{\sigma_s}{\sigma_t}\right|_{\text{elastic}} = \frac{4M^2 G_E^2}{-q^2 G_M^2}$$

(M = proton mass). Thus we expect G_E/G_M to approach a constant $(k_\perp^2 + m^2 \pm\Delta)_1/M^2$ for large q^2. A great problem is found when we estimate the constant from experiment. If $G_M \approx \mu G_E$ $\mu = 2.79$ we find $(k_\perp^2 + m^2 \pm \Delta)_1 = .11\ (\text{GeV})^2$ disturbingly small! But data at $q^2 = 2.6$ even shows G_E/G_M is less than 1, about .6 (and falling?) so $(k_\perp^2 + m^2 \pm \Delta)_1 < .04$?

Now we can see how f(x) should vary for x near 1, say $x = 1 - y$ with y small. Let us find the total probability that x exceeds $1 - y$ i.e. $\int_{1-y}^{1} f(x)dx$. We must again find one parton with nearly all the momentum, the total momentum of all the others cannot exceed y. That is not the same as the probability that none can exceed $x_0 = y$ which we worked out as $y^{-\gamma}$, but a little consideration of the problem soon shows that it does vary with the same power, but with a different constant in front. (This is most easily seen by a kind of dimensional analysis; or alternatively one can work out the case of a dx/x Poisson distribution in detail to check it.) Hence

$$\int_{1-y}^{1} f(x)dx \propto y^{\gamma}$$

for small y, or

$$f(x) \propto (1 - x)^{\gamma-1}$$

for x near 1. That is, we expect the power law of the elastic scattering to be related to the power law of the behavior or f(x) near x = 1. In particular we expect $f(x) \propto (1 - x)^3$ or a little higher. Data is not in disagreement with this. (This relation was first published by Drell and Yan.)

The region $-q^2$ large, M_x^2 finite

Resonances. Can we extend these ideas to get some understanding of why the resonances fall with about the same power of Q for large Q^2 as does the elastic peak? (Here we are studying the region large $-q^2$, $2M\nu \approx -q^2$, more precisely $M_x^2 = M^2 + 2M\nu - (-q^2)$ finite.) Perhaps it is better to frankly turn the question about the other way and ask what the fact that the resonances fall off as the elastic peak with $-q^2$ tells us about the wave function of these resonances.

If we excite say the Δ^+ our picture again cannot look like figure 29.1 for the "after" picture could not be a Δ^+ moving rapidly to the left because it contains some partons going in the wrong direction. We must instead have as in figure 29.2, but this time instead of asking whether the last picture looks like a proton, we ask: with what amplitude does it look like a Δ^+? We again expect a Δ^+ to have a sea region, etc., and to have a difficulty measured by a factor $Q^{-\gamma'/2}$ (in amplitude, for a proton $Q^{-\gamma/2}$) to appear with a single non-wee parton. We conclude from experiment the γ' probably equals γ - and can probably invent a posteriori argument to justify this. (For example, the sea may be very similar in the two cases because they are generated by the same parton - that is, the chance this parton is bare of non-wee particles is what the factor measures - and it is the same parton in either case.) The wee partons left over from the original right moving protons have only a certain amplitude to be in the correct proportions (of numbers non-Q dependent momenta, etc.) to be appropriate to a left-moving proton - and a different amplitude to be appropriate to a Δ^+. It is the ratio of these non-Q dependent amplitudes (squared) which determines the ultimate ratio of the probability of producing Δ^+ to proton at large Q^2. (It is always possible in principle of course that for some particular resonance this amplitude is zero and such a resonance would fall faster with Q than the elastic peak.)

These relations of power laws for elastic peak, resonances, and asymptotic f(x) are so interesting that it is important to notice that they can be expressed in a general principle (I am strongly indebted to J.D. Bjorken for discussions on this idea) quite apart from any parton theory interpretation. Stated in this way they are seen to be rather profound and represent a fundamental new property of high energy collisions (for corresponding relations are expected relating exclusive and inclusive hadron reactions at high energy).

Suppose we plot νW_2 for a given large q^2 against $M_x^2 - M^2 = 2M\nu(1-x) \approx -q^2(1-x)$ (x is near 1 and $2M\nu \approx -q^2$). We note the scale of $M_x^2 - M^2$ and 1-x are just proportional. We find a set of resonances at appropriate M_x^2

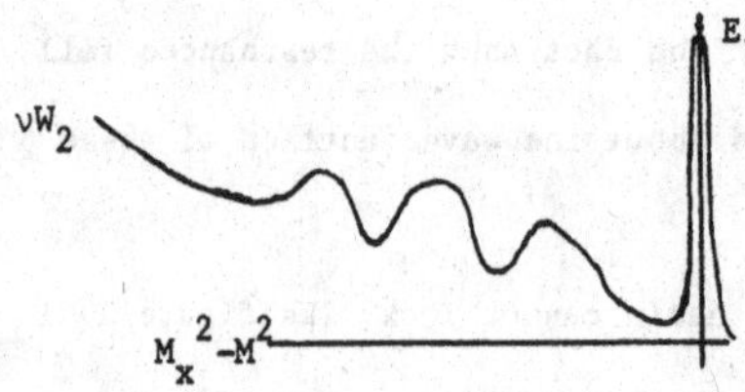

and the tail of the "background"curve $(1-x)^{\gamma-1}dx$ or $(-q^2)^{-\gamma}\,(M_x^2-M^2)^{\gamma-1}d(M_x^2)$. Now when we increase q^2 all the resonances fall (but stay in the same place M_x^2), as $(q^2)^{-\gamma}$ as we have seen, but the size of the "background" curve (at a given M_x^2) also falls as $(q^2)^{\gamma}$. That is the "background" (by background I mean the high x tail of the scaling region function $f_p(x)$) bears a fixed ratio to the resonances. It is not possible by going to higher q^2 (or in hadronic collisions to higher s) to separate even more clearly the resonances from the background, nor do the resonances fall into the background and get lost. Put another way, the background can be consistently viewed as due to overlapping resonances or tails of resonances itself and all resonances ultimately fall off the same way (as a power of q^2 as q^2 increases). Bohr would like this, he would say there are two complementary ways of viewing the background, as continuum single particle states leading to scaling expectations, or as large numbers of resonances - and nature conspires to make it impossible to decide in the intermediate region (large $M_x^2 - M^2$) which is correct.

Lecture 30

Argument that $\gamma' = \gamma$

We will make the assumption that for any state (e.g. p or Δ) as momentum Q increases the distribution among wee x (= finite momentum) approaches a definite distribution independent of Q. Rather, more explicitly for our case, we assume that the amplitude for the proton to be a single parton and a given distribution among wee momentum (e.g. below 2 GeV) and nothing in between falls somehow with Q but the distribution among the wees approaches a definite asymptote, (e.g. amplitude for no wee/amplitude for one of momentum 300 MeV/amplitude for two at 100 MeV and 400 MeV etc. become constant ratios),

$$\langle \text{one parton at } x \simeq 1, \text{ wees at momenta } k\text{'s} \mid \text{proton at } Q\rangle = F_p(Q) f_{p,i}(k\text{'s})$$

(the index i indicates parton type) and similarly for a resonance. Hence the scattering of a photon, which turns around the fast parton from Q to Q' (e.g. in a coordinate system where $q_\mu = (Q-Q', 0, 0, Q+Q')$).

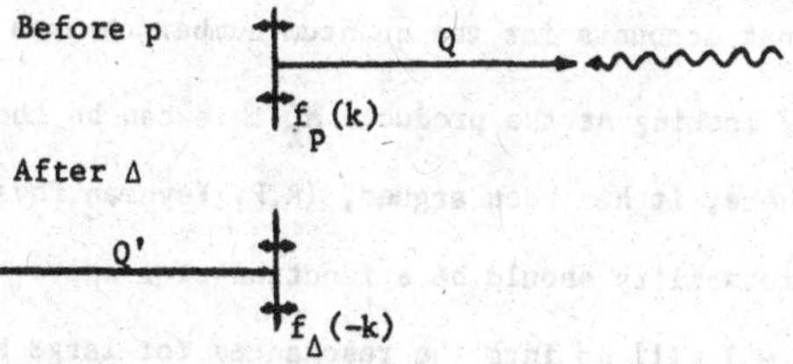

The amplitude is proportional to

$$F_p(Q)\ F_\Delta(Q') \int \bar{f}_p(k) f_\Delta(-k)\ d^3k$$

We explicitly suppose the last factor is not zero. (For certain resonances, or quantum number (e.g. I spin) reasons it might be zero, those may fall faster with Q, but we study the ones which are not special in this way.) The answer must be relativistically invariant, however, and thus be a function only of $q^2 = 4QQ'$. Hence $F_p(Q)\ F_\Delta(Q')$ is a function of QQ' only, from which we can deduce that both $F_p(Q)$ and $F_\Delta(Q)$ must be of the form $1/(2Q)^{\gamma/2}$ (a constant can be absorbed in the second factor).

Thus amp. $p \to \Delta = (q^2)^{-\gamma/2} a(p, \Delta)$

where

$$a(p, \Delta) = \sum_i \int \bar{f}_{p,i}(\vec{k}) f_{\Delta i}(-\vec{k})\ d^3k$$

depends on the states p, Δ, but not on q^2.

If two states have different powers of Q in their $F(Q) = Q^{-\gamma/2}$ then they must be orthogonal in the sense $\int \bar{f}_p(k) f_\Delta(-k)\, d^3k = 0$ - usually because of different quantum numbers, like charge strangeness or isospin (or z angular momentum) in the wee system.

The principle discussed at the end of the previous lecture very likely applies to high energy hadron collisions also; although not our direct interest in this course, I will briefly outline how it might work there.

Consider an exclusive reaction (at fixed or zero momentum transfer t) $A+B \to C$ + Resonance X, where A, B, C, are fixed. Let the incoming momenta in the c.m. be p and the outgoing momentum of $C = p_C = px$, $s = 4p^2$ goes to infinity; and imagine we measure the missing mass of the resonance by $M_X^2 = P_X^2 = (P_A + P_B - P_C)^2 \approx s(1-x) + M_A^2 + M_B^2 - M_C^2$. Thus plot resonance data as $M^2 = M_X^2 + M_C^2 - M_A^2 - M_B^2 \approx s(1-x)$. As s varies resonances stay at fixed M^2, of course, but their size goes down very likely as a power of s, say as $s^\alpha F(M^2)dM^2$ (α is negative); as expected for exclusive reactions, where α is the least negative possibility for an exchange of something that accounts for the quantum number changes from A to C. But since we are not really looking at the products M_X this can be thought of as an inclusive reaction where, it has been argued, (R.P. Feynman, Phys. Rev. Lett. 23 1415 (1969)) the probability should be a function of x only. Hence this scaling region near x = 1 will go into the resonances for large M^2 only if $s^\alpha F(M^2)dM^2 = s^\alpha F(s(1-x))ds(1-x)$ is a function of x only, not of s. Hence this function $F(s(1-x))$ must go as $(s(1-x))^{-\alpha-1}$ and the scaled result goes as $(1-x)^{-\alpha-1}dx$ for x near 1. (There are situations, for example, where C and A are both protons that it is technically hard to get to high enough x to avoid contamination of p's disintegrated from other hadron resonances C', etc., but these matters cannot concern us in detail here; the result has not been readily checked, therefore.)

Lecture 31

We summarize what we know or surmise about νW_2 and W_1 in the high ν region.

Region I) Large ν, large $-q^2$ but $-q^2/2M\nu = x$ finite. $2MW_1$ is a function of x only, called $f(x)$. $W_2 = (-q^2/\nu^2)W_1$.

Region II) Large ν, fixed $-q^2$ as $\nu \to \infty$.

$$2MW_1 = \frac{2M\nu}{-q^2} \cdot \text{a function of } q^2 = \frac{2M\nu}{-q^2} g(-q^2)$$

Region III) Large ν, large $-q^2$ but finite $M_x^2 - M^2 = 2M\nu - (-q^2)$.

$$2MW_1 = (2M\nu)^{-\gamma+1} \text{ function of } M_x^2 \text{ only} = (-q^2)^{-\gamma}\, h(M_x^2 - M^2)\, 2M\nu$$

(The last factor $2M\nu$ is the normalization from dM_x^2 to dx).

In order that the three regions fit together we have,

(I; II) for small x, $f(x)$ goes as a/x $(a = .32)$

for large $-q^2$, $g(-q^2)$ goes as a.

(II; III) for x near 1, $f(x)$ goes as $A(1-x)^{\gamma-1}$ $\gamma \approx 4$ or 5

for large $M_x^2 - M^2$, $h(M_x^2 - M^2)$ goes as $A(M_x^2 - M^2)^{\gamma-1}$

In region I we expect σ_s/σ_t to fall with increasing ν as $1/\nu$ or $1/(-q^2)$. In region II we expect σ_s/σ_t to approach some finite limit depending on $-q^2$ as $\nu \to \infty$. (This means νW_2 is finite, W_2 is of the same order as $(-q^2/\nu^2)W_1$ but is not equal to it.) In region III we expect σ_s/σ_t to fall with increasing ν as $1/\nu$. In the special region of small q^2: $W_2 = -q^2 W_1/\nu^2$ + order q^4/ν; $\sigma_s \approx q^2$.

General remark about the power law$(q^2)^{-\gamma}$, namely it may have a logarithmically falling coefficient $(q^2)^{-\gamma}/(a\ln q^2 + b)$. In the discussion of the previous lecture we saw powers where γ could depend on the quantum numbers of the wee group, like total isospin, angular momentum etc. These are discrete so the γ_i's are discrete and there is a lowest one. However, there is transverse momentum which is not discrete; γ could depend on the transverse momentum of the fast parton say as $\gamma = \gamma_0 + \gamma_1 k_\perp^2$ so the result is

$$\int (q^2)^{-(\gamma_0 + \gamma_1 k_\perp^2)}\, dk_\perp^2 = \frac{(q^2)^{-\gamma_0}}{\gamma_1 \ln q^2} \tag{31.1}$$

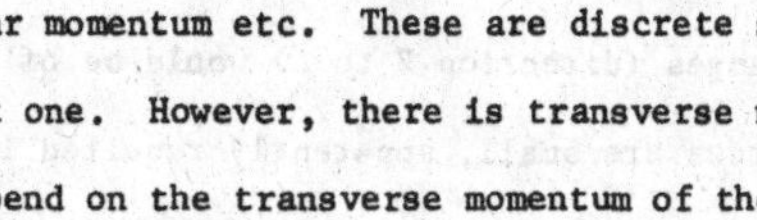

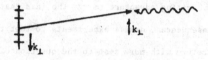

Partons as Quarks

Naturally we should like to go further and find out more about what quantum numbers the partons carry. As it turns out from present experiments only a few things can be learned, but future experiments can add information which greatly restricts the possibilities. We discuss this by choosing an example, that charged partons carry the quantum numbers of quarks. (In such a model Gell-Mann's equal time current commutation relations are automatically satisfied.) We are not making the three quark model, the number of quarks may be, indeed must be, infinite.

One is immediately struck with the question as to whether our explanation of scaling is possible at all if the partons do not have integer charges. The idea was that the outgoing parton (now quark) could escape without further large (as P) interaction to ultimately resolve itself into an outgoing bundle of hadrons.

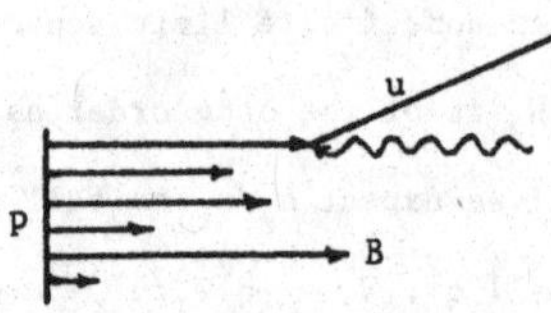

But it is a quark (knowing no low energy hadrons are of charge 2/3 say) it cannot turn into hadrons unless it picks up the extra 1/3 charge by dragging along another parton quark. If this extra parton is from the ongoing ones of momentum of order P the momentum changes (direction B to A) would be of order P and the assumption that binding forces are small, apparently required in our derivation, would fail. On the other hand, it is possible perhaps that the extra quarks are found among the wee partons, which are natural to any hadron wave function and since hadrons going in direction A or B share the same wee region they could exchange quarks there, making up the necessary integers. This is not clearly satisfactory but it is an exciting adventure to try the idea that charged partons are simply quarks, see consequences, devise experiments to test them; and if they are found to succeed, to return with more zest to the question of how Nature must then be resolving the apparent paradox among:

a. Partons carry quark quantum numbers

b. Hadrons do not

c. Scaling works.

In this spirit we look to discovering by what means we can check (a). At the very least we are giving an example of how experiment could lead to further identification of the character of partons.

To describe the parton distribution in a proton we would have six functions.

$u(x)$ = No. of up quarks with momentum x to x+dx <u>in the proton</u>

$d(x)$ = " down " " " " " "

$s(x)$ = " strange " " " " " "

Similarly $\bar{u}(x)$, $\bar{d}(x)$ and $\bar{s}(x)$ are the numbers of anti-up, anti-down and anti-strange quarks with momentum between x and x+dx in the proton.

The total charge on a proton is +1 so

$$1 = \frac{2}{3}\int_0^1 [u(x) - \bar{u}(x)]dx - \frac{1}{3}\int_0^1 [d(x) - \bar{d}(x)]dx - \frac{1}{3}\int_0^1 [s(x) - \bar{s}(x)]dx.$$

The z isospin is +1/2 so

$$\frac{1}{2} = \frac{1}{2}\int_0^1 [u(x) - \bar{u}(x)]dx - \frac{1}{2}\int_0^1 [d(x) - \bar{d}(x)]dx.$$

The strangeness is zero so

$$\int_0^1 [s(x) - \bar{s}(x)]dx = 0.$$

These equations have the solution

$$\int_0^1 [u(x) - \bar{u}(x)]dx = 2$$

$$\int_0^1 [d(x) - \bar{d}(x)]dx = 1$$

$$\int_0^1 [s(x) - \bar{s}(x)]dx = 0 \qquad (31.2)$$

That is, the net number of each kind of quark is just the number of that kind in the simple non-relativistic 3-quark model. The observed function $f_p(x)$ from $2MW_1$ (or $\nu W_2/x$) is the number of each kind weighed by the square of the charge: hence

$$f_p(x) = \frac{4}{9}\,(u(x) + \bar{u}(x)) + \frac{1}{9}\,(d(x)+\bar{d}(x)) + \frac{1}{9}\,(s(x)+\bar{s}(x)) \qquad (31.3)$$

Measurements on the ep scattering of course give us only this sum and we cannot extract the separate terms from this alone. On the other hand we have data on neutron scattering too. The neutron is given by the same formula except u(x) would now be the number of up-quarks in the <u>neutron</u>. However, by isospin reflection the neutron functions are got from those of the proton by replacing u by d, d by u. Thus, if u(x) continues to be as defined, the number of "ups" in the proton is also the number of downs in the neutron, we have

$$f_n(x) = \frac{1}{9}\,(u(x)+\bar{u}(x)) + \frac{4}{9}\,(d(x)+\bar{d}(x)) + \frac{1}{9}\,(s(x)+\bar{s}(x)) \qquad (31.4)$$

We have data on both these functions. The ratio f_n/f_p behaves like

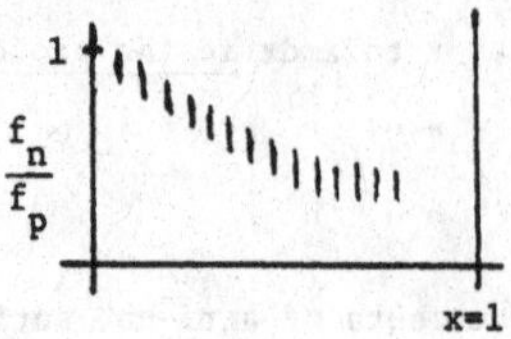

It has been suggested that possibly $f_n/f_p \sim 1-x$ falling to zero as $x \to 0$. We see immediately that this is impossible if charged partons are quarks; since $u(x)$, $\bar{u}(x)$ etc., are all positive f_n/f_p cannot fall below .25. The original data only went up to about $x \sim .8$ and had fallen to perhaps .3 so .25 was a possible limit as $x \to 1$ but we have since learned that there was a mistake in the computer program and it is closer to .4 for x near 1.

However, just for an exercise, what could we deduce if f_n/f_p was shown to be just 1/4 as $x \to 1$? That means $d(x)$ falls to zero faster than $u(x)$ as $x \to 1$. This would result if we supposed that the γ (in $(q^2)^{-\gamma}$ or $(1-x)^{\gamma-1}$) for wee partons of total isospin zero is smaller than the γ for isospin one (only one non-wee parton). Then the leading term (lowest γ) would require that the fast parton have the same isospin as the proton, up-quark only (since the wees are isospin zero). However, in that case the probability of exciting the Δ resonance (of isospin 3/2) would fall faster than the elastic peak for one can only make a Δ from wees with isospin one (as the leading quark can only contribute 1/2). Thus we would expect the Δ resonance to fall off with q^2 faster than others (data is insufficient to indicate whether this is true or false in fact).

We can argue (if partons are quarks) that if f_n/f_p falls below 1 it "almost surely" falls to 1/4 as follows: We first assume that almost surely (?) the values for γ for different quantum numbers among the wee are <u>not</u> equal and that one therefore dominates. First among the simpler ones of zero strangeness quarks we have γ_0 for $I = 0$ and γ_1 for $I = 1$. Hence depending on which is smaller we have eventually (as $x \to 1$) a pure $I = 0$ or pure $I = 1$ among the wees. For $I = 0$ the non-wee quark must be (1/2, 1/2) or u, this gives immediately f_n/f_p =

(charge of d)^2/(charge of u)^2 = 1/4. If I = 1 the non-wee quark can be u or d with weight given by the appropriate Clebsch-Gordan coefficient p = $\sqrt{2/3}$ (+1)d - $\sqrt{1/3}$ (0)u, n = $\sqrt{1/3}$ (0)d - $\sqrt{2/3}$ (-1)u therefore

$$\frac{f_n}{f_p} = \frac{(2/3)(4/9) + (1/3)(1/9)}{(2/3)(1/9) + (1/3)(4/9)} = \frac{3}{2} \qquad \text{if } I = 1.$$

Hence on the assumption of unequal γ's either 1/4 or 3/2 is the limit (therefore if below 1, it is 1/4). (There is the possibility that the non-wee quark is an antiquark, in that case if I = 0 dominates f_n/f_p = 2/3 but this possibility is considered very unlikely for p and n).

Another state may be wees of strangeness +1 and the non-wee pure s. It is unlikely this is easier than the non-strange case, but even so the ratio would be one, hence excluded.

Several models have been made which predict ratios below 1 but above 1/4. All of these assume the lack of correlation of fast quark quantum numbers and wee state character which we explicitly argued for, that is, they imply a degeneracy $\gamma_1 = \gamma_0$ in which case, of course, other values are possible.

The reader can show that the ratio of G_M for proton and neutron comes out -2 for isospin zero wees, for large $-q^2$ (it starts at μ_p/μ_n = -1.4 at q^2 = 0). The case that the total angular momentum among the wees is zero should be tried first. Show also that G_M for the proton is positive for large $-q^2$, so no sign change is necessary as q^2 rises, which is satisfactory because experimentally G_M for the proton never seems to fall to zero. Isospin one for the wees gives $G_M^p/G_M^n \to 0$ and G_M^n is positive at large $-q^2$.

Lecture 32

<u>Momentum Carried by the Quarks</u>

Another interesting study is the total momentum carried by each kind of quark.

Let $U = \int [u(x) + \bar{u}(x)]x\,dx$: $D = \int [d(x) + \bar{d}(x)]x\,dx$; $S = \int [s(x) + \bar{s}(x)]x\,dx$ be the total momentum carried by up (plus anti-up) quarks, etc. Then we know $\int x\, f_p(x)dx = .18$, $\int x\, f_n(x)dx = .12$ so

$$\frac{4}{9}U + \frac{1}{9}D + \frac{1}{9}S = .18$$

$$\frac{1}{9}U + \frac{4}{9}D + \frac{1}{9}S = .12 \qquad (32.1)$$

Now if we assume that there are no neutral partons beside quarks (e.g. no "gluons") all the momentum of the proton is quarks, we would have

$$U + D + S = 1$$

Solving these equations we have U = .21, D = .03, S = .76. This result is clearly unreasonable - that 3/4 of the momentum of the non-strange object proton is to be found in the strange quarks. This indicates most strongly that all partons cannot be quarks - there must be some neutral ones carrying momentum, say N. This momentum can be thought of as in the field, if any, by which the quarks interact; (possibly for example a neutral pseudovector) which if represented in field theory by intermediate field partons are usually called gluons.

In the region near $x \to 0$ the functions f_p and f_n diverge as $.32/x$, corresponding to infinite numbers of quarks and antiquarks. (The net number of quarks is finite, see eq. 31.2 and discussion.) Further f_p and f_n must have the same behavior so if $u(x) = \alpha/x$ for small x then $\overline{u}(x) = u(x) = d(x) = \overline{d}(x)$. We do not think we could prove $s(x)$ and its equal $\overline{s}(x)$ must equal that value also because SU(3) is not perfect, but it may not be far away (if indeed it is not equal).

An interesting quantity is the measure of how u and d approach each other as $x \to 0$; although each is infinite, the difference is finite and integrable. Experiment gives about .17 for the integral $\int_0^1 [f_p(x) - f_n(x)]dx$ but the error is large for it depends precisely on the small difference near x = 0 (as it is $\int_0^1 (\nu W^p{}_2 - \nu W^n{}_2)dx/x$.)

Models

Some people (as for exmaple Paschos, Weisskopf and Kuti) have tried to guess the functions u, d, etc., by making simplifying assumptions which are chosen for their simplicity rather than their physical necessity. Such predictions may or may not succeed, if they fail all we say is nature is not so simple. Unlike the ideas of the parton model we have used so far, the additional assumptions made in these models have in general no physical backing. In case any of you are interested we give an exmaple of such a theory here: It is that the wees or dx/x region is made entirely of uncorrelated pairs, and that on top of these there are three "valence quarks" of the kind of the three quark model and the wave function of these is a simple product of independent pieces. (We do not expect such simplicity.) That is, if $\alpha(x)$ is the typical sea distribution and $v(x)$ that of a valence quark we would write

$$u = \alpha + 2v \qquad \bar{u} = \alpha$$
$$d = \alpha + v \qquad \bar{d} = \alpha$$
$$s = \alpha \qquad \bar{s} = \alpha$$

for the proton. $\int_0^1 v(x)\,dx = 1$ (to get 31.2). Near $x \to 0$ it is $\alpha(x)$ that goes to infinity (as $.24/x$) but $x\,v(x) \to 0$ as $x \to 0$. This predicts for the total momenta carried by quarks $S = 2D - U$ so the data (32.1) gives $U = .36$, $D = .18$ and $S = 0$ (surprise) hence we are in some kind of trouble (α cannot be zero) unless the experimental errors allow $\int x f_n\,dx$ to exceed $\frac{2}{3}\int x f_p\,dx$.

At any rate some momentum (46%) must be in gluons. It predicts $\int_0^1 (f_p(x) - f_n(x))dx = \frac{1}{3}$ which is hard but not impossible to reconcile with the experimental .17. It also predicts $\nu W_{2n}/\nu W_{2p} \to 2/3$ (hence probably wrong) by supposing that as $x \to 1$, α falls faster than ν.

Lecture 33

Future tests of charged partons = quarks

So far no direct data is accurate enough to test or eliminate the idea that charged partons are quarks. However further experiments can. We give two examples, neutrino scattering and polarized electron and proton scattering.*

We shall discuss the inelastic scattering of neutrinos from proton $\nu p \to \mu$ plus anything in detail in the last part of this course. We do not know weak interaction theory as well as quantum electrodynamics but can use these experiments themselves to test it (e.g. if interaction is not a point-like four fermion interaction scaling will fail in the form derived below) - but supposing it is all right for now, we suppose the coupling with quarks is $(\bar{\mu}\,\gamma_\mu(1+i\gamma_5)\nu)(\bar{Q}'\gamma_\mu(1+i\gamma_5)Q)$ with the quark quantum numbers from $\bar{u}$ quark to $\cos\theta_c\,\bar{d} + \sin\theta_c\,\bar{s}$ where θ_c is Cabibbo's angle, $\sin^2\theta_c \approx .06$. For simplicity of our rough discussion here we take $\theta_c = 0$.

Then if we scatter $\bar{\nu}$ to μ^+ it can affect only up quarks (sending u to d) or $\bar{d}$ quarks ($\bar{d}$ to $\bar{u}$). Thus we would expect to measure just $u + \bar{d}$ if the current were pure vector, but it is V-A. The A is just like V for relativistic particles (except the sign is not reversed for antiparticles as it is for V.) If for the V-A current matrix element we write

$$M_{\mu\nu} = \sum \langle X|J^V_\nu - J^A_\nu|p\rangle^* \langle X|J^V_\mu - J^A_\mu|p\rangle \; 2\pi\delta(M_x^2 - M^2 - 2MV + q^2) \tag{33.1}$$

in analogy to electrodynamics we find that the general form by convention is

$$M_{\mu\nu} = \frac{\pi}{M}\,[4\,P_\mu P_\nu W_2 - 4M^2\delta_{\mu\nu}W_1 - 2i\varepsilon_{\mu\nu\sigma\lambda}\,P_\sigma q_\lambda W_3] \tag{33.2}$$

(leaving out all terms proportional to q_μ because they give small effects at the μ, ν end proportional to (mass $\mu)^2$). W_1, W_2 are as in electricity except they contain VV and AA (nearly equal); the new term W_3 comes from interference VA. In the scaling region νW_2, $2MW_1$, νW_3 scale to three functions $f_2(x)$, $f_1(x)$, $f_3(x)$ ($x = -q^2/2M\nu$). If partons have spin 1/2, $f_2 = 2\,x\,f_1$. For the quark model we get

$$\begin{aligned}
f_1^{\nu p} &= 2(\bar{u} + d\cos^2\theta + s\sin^2\theta)\\
f_1^{\nu n} &= 2(\bar{d} + u\cos^2\theta + s\sin^2\theta)\\
f_1^{\bar{\nu} p} &= 2(u + \bar{d}\cos^2\theta + \bar{s}\sin^2\theta)\\
f_1^{\bar{\nu} n} &= 2(d + \bar{u}\cos^2\theta + \bar{s}\sin^2\theta)\\
f_3^{\nu p} &= 2(\bar{u} - d\cos^2\theta - s\sin^2\theta)\\
f_3^{\nu n} &= 2(\bar{d} - u\cos^2\theta - s\sin^2\theta)\\
f_3^{\bar{\nu} p} &= -2(u - \bar{d}\cos^2\theta - \bar{s}\sin^2\theta)\\
f_3^{\bar{\nu} n} &= -2(d - \bar{u}\cos^2\theta - \bar{s}\sin^2\theta)
\end{aligned} \tag{33.3}$$

Thus we have many relations, for example

$$\begin{aligned}
f_1^{\nu p} - f_3^{\nu p} &= 4(d\cos^2\theta + s\sin^2\theta)\\
f_1^{\nu n} - f_3^{\nu n} &= 4(u\cos^2\theta + s\sin^2\theta)\\
f_1^{\bar{\nu} p} + f_3^{\bar{\nu} p} &= 4(\bar{d}\cos^2\theta + \bar{s}\sin^2\theta)\\
f_1^{\bar{\nu} n} + f_3^{\bar{\nu} n} &= 4(\bar{u}\cos^2\theta + \bar{s}\sin^2\theta)
\end{aligned} \tag{33.4}$$

---

* See also Appendix B

Neglecting $\sin^2\theta$

$$4\bar{u} = f_1^{\nu p} + f_3^{\nu p} \qquad 4u = f_1^{\bar{\nu}p} - f_3^{\bar{\nu}p}$$

$$4\bar{d} = f_1^{\nu n} + f_3^{\nu n} \qquad 4d = f_1^{\bar{\nu}n} - f_3^{\bar{\nu}n} \tag{33.5}$$

Sum rules

$$\int_0^1 (f_1^{\nu p} - f_1^{\nu p})dx = 2(2 - \cos^2\theta) \approx 2$$

$$\int_0^1 (f_1^{\bar{\nu}n} - f_1^{\nu n})dx = 2(1 - 2\cos^2\theta) \approx -2$$

$$\int_0^1 (f_3^{\nu p} + f_3^{\bar{\nu}p})dx = -2(-2 - \cos^2\theta) \approx -6$$

$$\int_0^1 (f_3^{\nu n} + f_3^{\bar{\nu}n})dx = -2(1 + 2\cos^2\theta) \approx -6 \tag{33.6}$$

the first is Adler's sum rule (not quark model dependent), the third was first derived by Llewellyn Smith and Gross.

Distribution predictions

$$f_3^{\bar{\nu}p} - f_3^{\nu p} = 2(-(u-\bar{u}) + \cos^2\theta\,(d-\bar{d}) + \sin^2\theta\,(s-\bar{s}))$$

$$= -6(f^{ep} - f^{en}) + \sin^2\theta\,((s-\bar{s}) - (d-\bar{d})) \tag{33.7}$$

the last term can probably be neglected, it gives a very close check on the model. Also

$$f_3^{\bar{\nu}n} - f_3^{\nu n} = 2(-(d-\bar{d}) + \cos^2\theta\,(u-\bar{u}) + \sin^2\theta\,(s-\bar{s}))$$

$$= 6\cos^2\theta_c\,(f^{ep} - f^{en}) + \sin^2\theta\,((s-\bar{s}) - (d-\bar{d})) \tag{33.8}$$

by subtraction with the previous relation can be used to eliminate the last term if there was any doubt.

Knowledge of $f_1^{\nu p}$, $f_1^{\bar{\nu}p}$, f^{ep}, f^{en} allows us to obtain $\bar{u} + \bar{u}$, $\bar{d} + d$, $\bar{s} + s$ separately and $f_3^{\nu p}$ permits further separation into u, $\bar{u}$, d, $\bar{d}$, $s + \bar{s}$ neglecting $\sin^2\theta$

<u>Deep inelastic scattering with spin</u>

If the electron beam and the proton is polarized in the deep inelastic scattering we can obtain still more information, this time about how the partons are polarized in a polarized proton. For definiteness we shall assume all the charged partons have spin 1/2.

For a particle at rest the state may be described by saying the spin is in some particular space direction $\bar{s}$. Relativistically transformed spin states can be described by a unit four-vector s_μ (the transformed s is as though it were a polar vector) satisfying $s_\mu s_\mu = 1$ and $s_\mu p_\mu = 0$ (p_μ = momentum of state). When taking matrix elements we can use spurs and sum over states if we replace the usual projection operator $\not{p} + m$ by the more complete $(\not{p}+m)\,(i-i\gamma_5\not{s})/2$ to project into the spin state s_μ. This s_μ is a pseudovector. The spin state of p in the J_ν factor of

$$W_{\mu\nu} = \sum \langle X|J_\nu|p'\rangle^* \langle X|J_\mu|p\rangle\ 2\pi\delta\,(M_x^2 - (P+q)^2)$$

need not be the same as the spin of p in the J_μ factor; but the quantity we want depends on spin as a 2 x 2 matrix (density matrix) one of two states in, one of two out. However all the information is contained if the diagonal is known for all states, not only for p = p' = +z or -z but also for p = p' = x = ((+z) + (-z))/$\sqrt{2}$ for example (in fact, these three and p = p' = -x are enough in our case). Thus we consider only diagonal cases p = spin p' = s_μ but for various s_μ. $W_{\mu\nu}$ can then be shown to be of the form

$$\frac{M}{\pi} W_{\mu\nu} = 4P_\mu P_\nu W_2 - 4M^2\delta_{\mu\nu}W_1 + 4Mi\varepsilon_{\mu\nu\lambda\sigma}q^\lambda[M^2 s^\sigma G_1 + (P\cdot q s^\sigma - s\cdot q P^\sigma)G_2] \tag{33.9}$$

(disregarding terms proportional to q_μ or q_ν) where W_1, W_2 are as before and G_1, G_2 are two new functions of q^2, ν.

What do we expect for these new functions G_1, G_2 in the deep inelastic region according to the parton model? If a parton has momentum $p_1 = xP$ and spin described by w^μ the scattering to $p_2 = p_1 + q$ is described by

$$sp\ ((\not{p}_2+m)\ \gamma_\mu\ (\not{p}_1+m)\ (1-i\gamma_5\not{w})\ \gamma_\nu)/2 =$$

$$2\ (p_{1\mu}p_{2\nu} + p_{1\nu}p_{2\nu} - \delta_{\mu\nu}(p_1\cdot q) + im\varepsilon_{\mu\nu\lambda\sigma}q^\lambda w^\sigma) \tag{33.10}$$

Let us use for example the coordinate system where the photon has pure z component

$$q = (0,\ -2P\ x,\ 0,\ 0) \qquad 2M\nu = 4p^2 x$$

$$P_\mu = (\varepsilon,\ P,\ 0,\ 0) \qquad -q^2 = 4P^2\,x^2$$

Now first suppose the helicity of the proton is +, thus $s_\mu = (P,\ \varepsilon,\ 0,\ 0)/M$.

Further let

$h_+(x)$ = No. of partons · $(\text{charge})^2$ with helicity + in a proton of hel. +

$h_-(x)$ = No. of partons · $(\text{charge})^2$ with helicity − in a proton of hel. −

we have $w^\mu = (P_1, \varepsilon_1, 0, 0)/m = xMS^\mu/m$ for hel. + and $w^\mu = xMS^\mu/m$ for hel. −.

We therefore have

$$\frac{1}{2\pi} W_{\mu\nu} = \int \frac{1}{\xi} (h_+ + h_-)(4\xi^2 P_\mu P_\nu - 2\xi\delta_{\mu\nu} P\cdot q)\ \delta\left((\xi P+q)^2 - m^2 - \Delta\right) +$$

$$+ \int \frac{1}{\xi}(h_+ - h_-)(2im\varepsilon_{\mu\nu\lambda\sigma} q^\lambda w^\sigma)\ \delta\left((\xi P-q)^2 - m^2 - \Delta\right) \qquad (33.11)$$

The δ function gives $\delta(\xi - x)/2M\nu$ in the scaling limit. The first integral gives the results we already know $2MW_1 = f(x) = h_+ + h_-$ and $\nu W_2 = xf(x)$. The second integral gives

$$2M^2\nu\,(G_1 - 2xG_2) = h_+(x) - h_-(x) \qquad (33.12)$$

therefore the function $G_1 - 2\,x\,G_2$ scales.

Suppose now that the proton is polarized in + x direction, we have $s^\sigma = w^\sigma = (0, 0, 1, 0)$ for parton with pol. + x and $w^\sigma = -s^\sigma$ for pol. −x.

$k_+(x)$ = No. of partons · $(\text{charge})^2$ with pol. + x in a proton of pol. + x

$k_-(x)$ = No. of partons · $(\text{charge})^2$ with pol. − x in a proton of pol. − x.

We have an expression of the type (33.11) with h_+ replaced by k_+ and h_- replaced by k_- from which we derive

$$2M^3\nu G_1 + 2M^2\nu^2 G_2 = m(k_+(x) - k_-(x))/x \qquad (33.13)$$

therefore the function $2M^3\nu G_1 + 2M^2\nu^2 G_2$ scales.

Hence from (33.12) and (33.13) $M^2\nu^2 G_2$ scales and $M\nu^3\,G_1$ scales and we write $M^2\nu G_1 = g_1(x)$, $M\nu^2 G_2 = g_2(x)$. These are the scaling predictions for spin cases. The term in G_2 in (33.12) is relatively negligible (order $1/\nu$ times term in G_1). We have

$$2\,g_1(x) = h_+(x) - h_-(x) \qquad (33.14)$$

$$2(g_2(x) + g_1(x)) = k_+(x) - k_-(x) \qquad (33.15)$$

$$h_+ + h_- = k_+ + k_- = f(x)$$

Of course, these new functions g_1, g_2 cannot be obtained from f(x) because new distributions are involved, the number of partons with various spins is states of various helicity, so that we derive nothing but the scaling rules even if we use a special model like quarks, etc. However, in that model we can obtain a remarkable sum rule of Bjorken.

We can use our proton wave function (described by parton) not only to obtain predictions for deep inelastic scaling but like any other wave function we can write the expectation value of various operators for it. We have already done that trivially when we wrote j_μ as the operator $\sum_{\text{quarks}} e_i\gamma_\mu$ and found the total charge on the proton to be expressed as $\int_0^1\left[\frac{2}{3}(u-\bar{u}) - \frac{1}{3}(d-\bar{d}) - \frac{1}{3}(s-\bar{s})\right] dx$.

Now we will discuss the expectation of a less trivial operator, the axial β decay operator for $p \to n$ transitions. It is written empirically as $\langle p|J^+_{A\mu}|n\rangle = i(\bar{u}_p\gamma_\mu\gamma_5 u_n)(G_A/G_V)$ and experimentally $G_A/G_V = 1.23 \pm .02$. Theoretically in the quark model the operator $J^+_{A\mu}$ is the sum over quarks of $\tau^+\gamma_\mu\gamma_5$ (τ^+ I spin raising op). If we change the I spin to third component we can use the relation

$$\langle p|J^3_{A\mu}|p\rangle = \langle p \sum_{\text{quarks } i} \tau^3\gamma_\mu\gamma_5|p\rangle = \bar{u}_p\gamma_\mu\gamma_5 u_p(G_A/G_V)(1/2)$$

We calculate both sides for a proton moving fast to the right with momentum P and spin s^σ and sum on proton states via a spur operator correctly projecting. On the left side we have the corresponding sum on quarks, thus

Total of sp $[(\not{p}+m)(1-i\gamma_w\not{w})\gamma_5\gamma_\mu] = \frac{1}{2}(G_A/G_V)$ sp $[(\not{P}+M)(1-i\gamma_5\not{s})\gamma_5\gamma_\mu] = \sum mw^\mu$ summed with weight + 1/2 for u, $\bar{u}$; − 1/2 for d, $\bar{d} = (G_A/G_V)s^\sigma M/2$. Now the sum on the left is exactly the integral over x of 2 g_1, (or $2(g_1+g_2)$ depending on the spin case) that we needed for deep inelastic ep scattering except the weights in the latter case were 4/9 for u, $\bar{u}$,1/9 for d, $\bar{d}$ and 1/9 for s, $\bar{s}$ for the proton (or 1/9 for u, $\bar{u}$, 4/9 for d, $\bar{d}$, 1/9 for s, $\bar{s}$ for the neutron) while here we want 1/2 for u, $\bar{u}$, − 1/2 for d, $\bar{d}$. But the difference of g_1 for proton g_{1p} and neutron g_{1n} has weights 1/3 for u, $\bar{u}$, − 1/3 for d, $\bar{d}$ or 2/3 of what we want. Hence we have for the two polarization cases

$$2\int_0^1 (g_{1p} - g_{1n})dx = \frac{1}{3}\left(\frac{G_A}{G_V}\right) \qquad (33.16)$$

$$2\int_0^1 (g_{1p} - g_{1n} + g_{2p} - g_{2n})dx = \frac{1}{3}\left(\frac{G_A}{G_V}\right) \qquad (33.17)$$

((33.16) is Bjorken's relation). From these we conclude $\int_0^1 (g_{2p} - g_{2n})dx = 0$. This comes merely from the equality of the coefficient for the polarization z and x cases, an equality which would have to hold no matter what the weights were; so we can conclude individually $\int_0^1 g_{2p}dx = 0$ and $\int_0^1 g_{2n}dx = 0$, i.e.

$$\int_0^1 g_2 dx = 0 \tag{33.18}$$

a result of rotational symmetry (angular momentum conservation).

The first relation is more remarkable and would test simultaneously the quark view, and the interpretation of the weak current as $\gamma_\mu(1+i\nu_5)$ not only for the elementary particles e , ν_e, μ, ν_μ but also for the elementary constituent quarks (or quark quantum numbers on spin 1/2 partons). Its verification, or failure, would have a most decisive effect on the direction of future high energy theoretical physics.

Tests of the Parton Model

Lecture 34

Angular momentum in parton wave functions

The equation $\int_0^1 g_2 dx = 0$ comes from angular momentum considerations (that the proton is spin 1/2 and the parton is spin 1/2). This brings up a general research problem about angular momentum and parton wave functions (functions of x and transverse momentum $k_\perp$). The dependence of these functions on the transverse momenta $k_\perp$ is not known, and are very interesting theoretically. Can any information on this, or other questions, be got by studying the angular momentum properties of parton wave functions? For example, what restrictions if any on the wave function come from the fact that the total angular momentum of the proton is just 1/2? Or, suppose the parton probability distribution f(x) were known for a Δ of helicity + 3/2. What could you say about the distribution for a Δ of helicity + 1/2?

Other experiments testing parton idea (Drell)

Drell has suggested another experiment which can be predicted by knowledge of the parton distribution. Again we use the parton as quark model to describe the idea, but it can of course be analyzed the same way in other representations for partons. The reaction is $p + p \to \mu^+\mu^- +$ anything. Proton plus proton (or $p + \bar{p}$) at very high s (e.g. P in CM is large) makes a $\mu^+\mu^-$ pair whose energy is proportional to P, and whose z component of momentum is of order P, as $P \to \infty$. The picture is

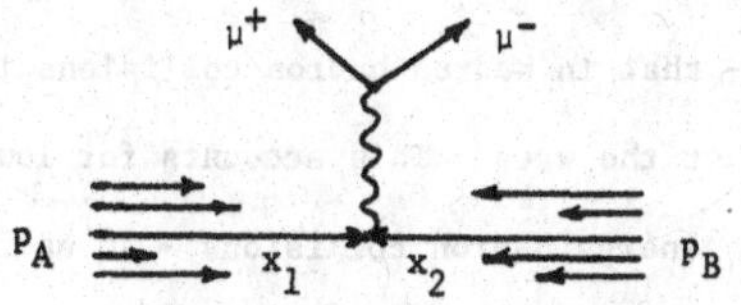

the μ pair is made from the annihilation of a parton in proton A and anti-parton in proton B and the other partons just keep going on to make final hadrons, the "anything". By suitable choice of the energy and momentum (in z direction) of the μ's we can determine x_1 and x_2 of the annihilating partons. (Thus if, in the CM, the momentum of the colliding protons is P, and the energy and total z component momentum of the μ's is ε, p_z, we have by conservation of momentum and energy

$$Px_1 - Px_2 = p_z \quad ; \quad Px_1 + Px_2 = \varepsilon$$

$$Px_1 = (\varepsilon+p_z)/2P \quad , \quad x_2 = (\varepsilon-p_z)/2P$$

It will be easiest to measure the pair at large angles (the angular distribution is $1 + \cos^2\theta$) so no great loss of intensity occurs here, and background hadrons which should move primarily in z direction are avoided. If $\theta = 0$ is chosen however, the momenta of the two μ's give directly Px_1, Px_2 because in such a one-dimentional collision of relativistic particles the momentum of the final particles is the same as that of the incident particles.

The cross section for this is the cross section for relativistic pair annihilation ($e^4(1 + \cos^2\theta)/16\ \varepsilon^2$) times the chance C of finding the appropriate partons to annihilate; e.g. for p + p, if partons are quarks:

$$C = \frac{4}{9}\,(u(x_1)\bar{u}(x_2)+\bar{u}(x_1)u(x_2)) + \frac{1}{9}\,(d(x_1)\bar{d}(x_2)+\bar{d}(x_1)d(x_2)) + \\ + \frac{1}{9}\,(s(x_1)\bar{s}(x_2)+\bar{s}(x_1)s(x_2)) \qquad (34.1)$$

To be valid, of course, the wee region must be avoided (that is, P must be high enough that Px_1, Px_2 are large compared to 1 GeV). One might at first be concerned that the $|P\rangle|P\rangle$ state we start with is not a stationary state, for protons scatter via strong interactions so the possibility exists that the parton distributions are first disturbed by "strong" interactions before they annihilate to make μ's. But we are coming to understand that the "strong interactions" are not so strong in this sense - that in hadron-hadron collisions it is not the fast partons that interact, but the wees. This accounts for low momentum transfer and scaling in very high energy hadron collisions - so as long as we avoid this wee interacting region (interactions via finite energies, not of order P) the partons act as free particles nearly (as $P \to \infty$).

Obviously with enough data we could see if a form like (34.1) could work - or if we know u, $\bar{u}$ etc. from neutrino scattering we could test it directly.

Lecture 35

$p + p \to \mu^+\mu^-$ + anything (continued)

Gronau has pointed out that with no further data than we already have we could calculate this factor C in at least one region. If x_1 is small, we know all the functions u, $\bar{u}$, d, $\bar{d}$ go as α/x_1 with the same α, and s, $\bar{s}$ go as α_s/x_1, where α_s may be close to α (but not necessarily exactly equal, say $\alpha_s = \alpha - \beta$ with β small; SU_3 says $\alpha_s = \alpha$ but it is incorrect). We shall first consider $\beta = 0$, and then say what the effect of it is. Hence in this region ($\alpha = .24$) we have

$$C = \frac{\alpha}{x_1}\left[\frac{4}{9}(u(x_2) + \bar{u}(x_2)) + \frac{1}{9}(d(x_2) + \bar{d}(x_2)) + \frac{1}{9}(s(x_2) + \bar{s}(x_2))\right] \quad (35.1)$$

But the last factor is just $f^{ep}(x)$, the $2MW_1$ for the proton and is known, so C is predictable in this region. It shows that Drell's experiment could easily serve as a test of any parton model - and will be an interesting experiment to watch.

If β is not zero we get

$$C = \frac{.24}{(1-\beta/6)x_1}\left[f^{ep}(x_2) - \frac{\beta}{9}(s(x_2) + \bar{s}(x_2))\right]$$

so it seems that there is little effect from any reasonable β. For example, for x_1, x_2 small

$$C = \frac{4}{3}\,\frac{(.24)^2}{x_1x_2}\left[1+5\left(\frac{\beta}{6-\beta}\right)^2\right]$$

a very weak dependence on β.

In a similar way we could describe the proton antiproton case $p+\bar{p}\to\mu^+\mu^-$ + anything. It would be determined by a factor

$$C' = \frac{4}{9}\,(u(x_1)u(x_2)+\bar{u}(x_1)\bar{u}(x_2)) + \frac{1}{9}\,(d(x_1)d(x_2)+\bar{d}(x_1)\bar{d}(x_2)) + \\ + \frac{1}{9}\,(s(x_1)s(x_2)+\bar{s}(x_1)\bar{s}(x_2)). \tag{35.2}$$

If finally, the functions u, $\bar{u}$ etc. are determined for the proton, the corresponding functions for the pion $u_x(x)$, $\bar{u}_x(x)$ etc. could be obtained through (difficult to get enough energy) $\pi + p \to \mu^+\mu^- + x$ for now

$$C_\pi = \frac{4}{9}\,(u_p(x_1)\bar{u}_\pi(x_2)+\bar{u}_p(x_1)u_\pi(x_2)) + \frac{1}{9}\,(d_p(x_1)\bar{d}_\pi(x_2)+\bar{d}_p(x_1)d_\pi(x_2)) + \\ + \frac{1}{9}\,(s_p(x_1)\bar{s}_\pi(x_2)+\bar{s}_p(x_1)s_\pi(x_2)) \tag{35.3}$$

<u>Electron pair production of hadrons</u>

Drell has suggested that another fundamental experiment may yield information about the character of partons. It is $e^+ + e^- \to$ any hadrons at high energy. We would expect to produce some spin 1/2 parton pair of charge e_i,

$$\frac{d\sigma}{d\Omega} = e_i^{\,2}\,\frac{e^4}{4q^2}\,(1+\cos^2\theta) \quad , \quad \sigma = e_i^{\,2}\,\frac{4\pi e^4}{3q^2} \tag{35.4}$$

$= (e_i/e)^2\sigma_o$ where σ_o is the cross section for spin 1/2 pair production (e.g. $e^+e^-\to\mu^+\mu^-$). Now this pair would turn into hadrons - and if the energy is high enough, of nearly the same energy as the virtual state of a parton pair. Thus if there were a number of parton pair types of charges e_1, e_2 etc., the total cross section $\sigma_{e^+ + e^- \to}$ hadrons is just the total to make each pair:

$$\frac{\sigma(e^+e^- \to \text{hadrons})}{\sigma(e^+e^- \to \mu^+\mu^-)} = \sum_i e_i^{\,2} \tag{35.5}$$

That is, the ratio approaches a constant, and the constant is the sum of squares of parton charge summed on each type that exists. If there are any spin 0 partons they would contribute $\frac{1}{4}$ as strongly (e.g. $\frac{1}{4}e^2$ for each spin zero parton). We call $D = \sum_i e_i^{\,2}$ "Drell's constant."

For example if partons are quarks we expect $\sum e_i^{\,2} = \frac{4}{9} + \frac{1}{9} + \frac{1}{9} = 2/3$ for D. If partons must carry integral charges, and be spin 1/2, the sum must be at least

1. In the case of quarks we would have the difficulties we have disregarded as to how a pair of oppositely moving charges of $\pm$ 2/3 say can turn into only hadrons of integral charge. There may be some way to get the required extra quarks from a soft sea of pairs, but one can perhaps legitimately question whether the above arguments are entirely clear, involving as they do an assumption that the parton states go entirely into hadron states. Is it really consistent to do this and omit states of quark quantum numbers? It would be most exciting if the ratio does come out 2/3 as well it might, for the theoretical questions produced by such a simple answer would be very interesting. I think it would be worthwhile as an exercise beforehand to assume the result was really 2/3 and see what paradoxes, if any, would then have to be resolved.

As we discussed previously theoretically (lecture 5), the total cross section for $\sigma(e^+e^- \to$ any hadrons) as a function of q^2 ($q^2 = 4E^2$, E = CM energy) determines the vacuum expectation value of the product of currents

$$<0|J_\mu(-q)J_\nu(q)|0> = (q_\mu q_\nu - \delta_{\mu\nu} q^2)\,\theta(q_0)p(q^2) \tag{35.6}$$

with

$$\sigma = \frac{(4\pi e^2)^2}{2q^2}\,p(q^2)$$

The vacuum scattering of virtual photons by virtual hadrons is determined by

$$\text{F.T.} <0|\{J_\mu(1)J_\nu(2)\}_T|0> = (q_\mu q_\nu - q^2\delta_{\mu\nu})v(q^2) \tag{35.7}$$

where dispersion theory told us $\text{Im}(iv) = \frac{1}{2}\,p(q^2)$,

$$4\pi e^2 i[v(q^2)-v(0)] = \frac{q^2}{\pi(4\pi e^2)}\int_{4m_\pi^2}^{\infty}\frac{\sigma(s)ds}{s-q^2} \tag{35.8}$$

(1 + above quantity) is the factor by which a photon propagator $1/q^2$ must be multiplied for first order effect of virtual hadrons i.e.

$$= \frac{1}{q^2}\,4\pi e^2{}_i[v(q^2)-v(0)]$$

The low energy tests, like effect on Lamb shift depend just on the lowest q^2 or only on iv'(0) that is $\int \sigma s^{-1} ds$.

Today we can begin to say something quite detailed about this, for in the region $(2m_\pi)^2$ to just above 1 GeV^2 the cross section is dominated by the ρ, ω, ϕ production. Perhaps $p(q^2)$ settles down to $D/6\pi$ for large q^2 fairly soon. Although one might suppose we might have to go decidedly above the nucleon pair creation, this may not be true; soft particle production may dominate at all energies and reach its asymptote sooner.

A. Cisneros has calculated the various hadronic contributions to $iv'(0)$. The contributions from the ρ, ω and ϕ turn out to be ten times larger than the large q^2 contribution under reasonable assumptions about the likely value of Drell's constant and the value of q^2 at which $p(q^2)$ attains its asymptote.

Data from e^+e^- intersecting rings can be used directly to obtain the contribution from the ρ. The data is simply inserted in the formula

$$iv'(0) = \frac{1}{\pi(4\pi e^2)^2}\int_{4m\pi^2}^{\infty}\frac{\sigma(q^2)}{q^2}\,dq^2 \tag{35.9}$$

to obtain

$$iv'_\rho(0) = (5.5 \pm .5)10^{-2}\ \text{GeV}^{-2} \tag{35.10}$$

(What was used here was the data on the pion form factor which is almost everything there is in $\gamma(\text{virtual}) \to \rho \to$ hadrons. The 4π inelasticity is very small even at $q^2 = (.8\ \text{GeV})^2$).

In the case of the ω and ϕ we assume VDM, which works well in this type of process. In this case the contribution from the vector meson $v = \omega$ or ϕ is

$$p(q^2) = \frac{2}{g_v^2}\frac{\Gamma_v}{m_v}\frac{m_v^4}{(q^2-m_v^2)^2+\Gamma_v^2 m_v^2} \tag{35.11}$$

From the relation $\text{Im}\,(iv(q^2)) = \frac{1}{2}p(q^2)$ we deduce

$$iv_v(q^2) = \frac{1}{g_v^2}\frac{m_v^2}{(q^2-m_v^2)-i\Gamma_v m_v} \tag{35.12}$$

For a very narrow resonance, which is a good approximation for ω and ϕ, we have $p(q^2) = m_v^2 g_v^{-2} 2\pi\delta(q^2-m_v^2)$ the value of $iv'(0)$ is simply

$$iv_v'(0) = g_v^{-2} m_v^{-2} \tag{35.13}$$

(This formula gives $iv'_\rho(0) = 5.3 \times 10^{-2}\ \text{GeV}^{-2}$ for the ρ, in good agreement with what was obtained above using the data directly.) The contributions of the ω and ϕ are therefore

$$iv'_\omega(0) = (.7 \pm .1)\ 10^{-2}\ \text{GeV}^{-2} \tag{35.14}$$

$$iv'_\phi(0) = (.57 \pm .06)\ 10^{-2}\ \text{GeV}^{-2} \tag{35.15}$$

We now evaluate the contribution from the "tail" of $p(q^2)$ assuming it has attained its asymptotic value $D/6\pi$ at $q^2 = q_o^2$

$$iv'_t(0) = \frac{1}{2\pi}\int_{q_o^2}^{\infty} \frac{D}{6\pi} \frac{dq^2}{q^4} = \frac{D}{12\pi^2} \frac{1}{q_o^2} \qquad (35.16)$$

This gives $iv'_t(0) = .84 \times 10^{-2}\ Dq_o^{-2}$. If q_o^2 is as low as 1 GeV^2 and D is the quark value 2/3 we have $.56 \times 10^{-2}\ GeV^{-2}$ for $iv'_t(0)$. We believe the actual value of the non-vector meson contribution not to be much larger than this, we assign to it 100% uncertainty. Adding the various contributions we get finally

$$iv'(0)_{\text{hadrons}} = (7.3 \pm 1.1) \times 10^{-2}\ GeV^{-2}$$

This corresponds to a correction to the magnetic moment of the muons of $(5.5 \pm .7) \times 10^{-8}$ in $(g-2)/2$.

For comparison, $iv'(0)$ for muons is $[4\pi^2\ 15\ M_\mu^{\ 2}]^{-1} = 15.3 \times 10^{-2}\ GeV^{-2}$, so the hadronic contribution is about one half as large.

Away from $q^2 \approx 0$, the contributions to $i(v(q^2) - (0))$ from the tail of $p(q^2)$ grow as the vector meson contributions fall, so the uncertainties are greater.

Another expectation of interest from the parton model is the angular distribution of hadronic products in high energy e^+e^- collisions.

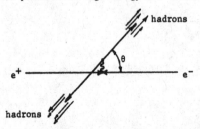

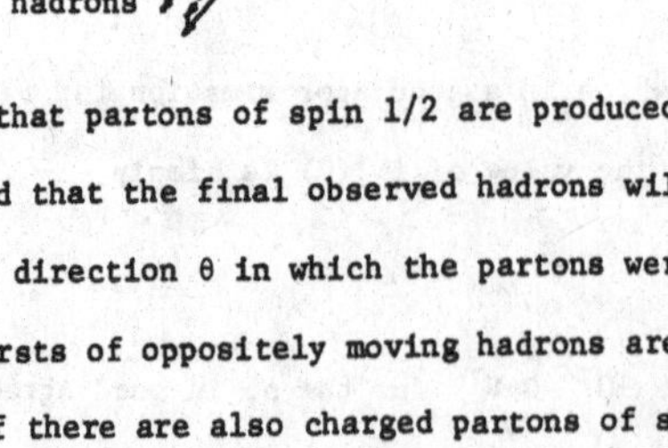

We expect that partons of spin 1/2 are produced with an $(1+\cos^2\theta)$ angular distribution and that the final observed hadrons will have small transverse momenta relative to the direction θ in which the partons were produced at high enough energy. Two bursts of oppositely moving hadrons are expected which will determine the angle θ. If there are also charged partons of spin 0 the angular distribution will be

$$D_{1/2}\ (1 + \cos^2\theta) + \frac{1}{2}\, D_o\ (1 - \cos^2\theta)$$

where $D_{1/2}$ and D_o are Drell's constants for partons of spin 1/2 and spin 0 respectively. For quarks we expect of course only a $(1 + \cos^2\theta)$ distribution.

Inelastic Scattering as Properties of Operators

Lecture 36

Inelastic e p scattering as properties of operators

We should now like to turn back to a point of view, described more particularly in the first lectures of this course, of the electrodynamic properties of hadrons as being measurements of properties of the current operators $J_\mu(1)$. For example there we said we expected that the commutator $[J_\nu(2), J_\mu(1)]$ would vanish if 1, 2 were outside each others light cone. Second order interactions were described by matrix elements of a second operator $V_{\nu\mu}(2, 1)$ which we showed would be expressible as a time ordered product of the first ones J_μ (written $\{J_\nu(2)\ J_\mu(1)\}_T$).

As we have seen the e p scattering measures a function which we have written as

$$K_{\nu\mu}(q) = \sum_x \langle p|J_\nu(-q)|x\rangle\langle x|J_\mu(q)|p\rangle\ 2\pi\delta\ (M_x^2-(p+q)^2).$$

We now consider it in a more abstract way. First we note that if the states p and x are imagined to contain their center of mass momentum factors, the $2\pi\delta(M_x^2-(p+q)^2)$would be unnecessary, for the class of states x' now mean one of mass M_x moving with momentum P. Thus

$$K_{\nu\mu}(q) = \sum_{\text{all } x'} \langle p|J_\nu(-q)|x'\rangle\langle x'|J_\mu(q)|p\rangle$$

or by completeness,

$$K_{\mu\nu}(q) = \langle p|J_\nu(-q)J_\mu(q)|p\rangle$$

Thus $K_{\nu\mu}$ is the expectation on the proton of the product of two operators.

We have measured it only for $q^2 < 0$ but it exists also for $q^2 > 0$ and of course the entire function for all q^2 both positive and negative belongs together theoretically. For positive q^2, however, there is a small technical point. We wish to alter the definition because we do not wish diagrams which do not affect the proton at all, thus

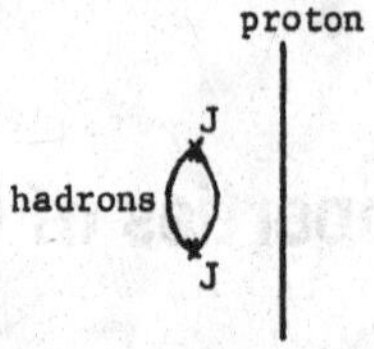

Such terms do not exist for the non-diagonal matrix elements of the product of operators term $\langle p'|J_\nu(-q')J_\mu(q)|p\rangle$ and we should like to define K as the limit of this as $p' \to p$ to obtain a suitably useful definition of matrix elements of the product operator. We can get the same result if we explicitly subtract the contributions of the disconnected diagrams of the type above. Hence we write

$$K_{\mu\nu}(q) = \langle p|J_\nu(-q)J_\mu(q)|p\rangle - \langle 0|J_\nu(-q)J_\mu(q)|0\rangle\langle p|p\rangle \tag{36.1}$$

Then if the final p is changed to p' slightly different, the expression defined by (36.1) is continuous.

We would like to consider, for theoretical purposes, various other currents beside electromagnetic – we can have various SU_3 generalizations in an octet (or nonet) (i.e. currents with different charge numbers) and axial currents as well. Thus we let these indices (SU_3 and axial) as well as the μ, ν indices of space be contained in a single letter A (calling $J_\mu(1) = A(1)$ etc.) to save writing in general arguments, and can always go back and insert indices at the end. Thus if A is any operator of our allowed set (for our examples, a vector or axial vector local current in SU_3 octet or possibly singlet) we write (obviously the diagonal proton state can be generalized immediately too, but we leave that for the student).

$$K_{BA}(q) = \langle p|B(-q)A(q)|p\rangle - \langle 0|B(-q)A(q)|0\rangle\langle p|p\rangle \tag{36.2}$$

And lastly we omit writing the last term for simplicity, but you must remember that, for diagonal matrix elements it is there.

K_{BA} is the diagonal matrix element of the product of two operators. Evidently we can define a corresponding thing in space:

$$K_{BA}(2,\ 1) = \langle p|B(2)A(1)|p\rangle$$

Since the diagonal element K depends only on the difference $x_{2\mu} - x_{1\mu}$ or 2-1, its Fourier transform is $K_{BA}(q)$. We explicitly call the components of $q = (\nu,\ \vec{Q})$ in the system where the proton is at rest (others via relativistic transformation). Thus $|p\rangle$ is at rest

$$K_{BA}(\nu,\ \vec{Q}) = \langle p|B(-\nu,-\vec{Q})A(\nu,\vec{Q})|p\rangle$$

$$= \sum_{\text{all } x} \langle p|B(-\vec{Q})\ x\rangle \frac{1}{2E_x} \langle x|A(\vec{Q})|p\rangle \delta(E_x - E_p - \nu) \qquad (36.3)$$

Since K gives the product operator we can find from it the commutator as well as the time ordered operator so useful in scattering.

We note, in our case (in the rest system of the proton), since the proton is the lowest mass of all states of baryon number one (and A does not change baryon number) that $E_x > E_p$ for all x. Hence if $\nu < 0$ the δ function is zero - no state can be reached by $A(\nu,\ \vec{Q})|p\rangle$ if $\nu < 0$ - and ($|p\rangle$ at rest)

$$K_{BA}(\nu,\ \vec{Q}) = 0 \text{ if } \nu < 0\ . \qquad (36.4)$$

Now the commutator matrix element is defined by

$$C_{BA}(2,1) = \langle p|[B(2),\ A(1)]|p\rangle = K_{BA}(2,\ 1) - K_{AB}(1,\ 2)$$

so its Fourier transform satisfies

$$C_{BA}(\nu,\ \vec{Q}) = K_{BA}(\nu,\ \vec{Q}) - K_{AB}(-\nu,\ -\vec{Q}) \qquad (36.5)$$

We note that K can be obtained from C and vice versa because of (36.4). Thus

$$K_{BA}(\nu,\ \vec{Q}) = C_{BA}(\nu,\ \vec{Q}) \text{ for } \nu > 0 \qquad (36.6)$$

and $C_{BA}(-\nu,\ -\vec{Q}) = -C_{AB}(\nu,\ \vec{Q})$.

Thus in measuring $K_{BA}(\nu,\ Q)$ we are measuring the Fourier transform of the commutator of two currents.

We shall discuss the consequences of this interesting result in a moment, but while we have these equations before us we wish to derive a few formulas for the scattering amplitude which we will need later in the course. As we have discussed the scattering amplitude for an incoming photon (virtual or real) coupled to J (1) say A(1) to an outgoing one coupled to B is determined by the operator

$$V_{BA}(2, 1) = \{B(2)A(1)\}_T + \text{seagulls} \tag{36.7}$$

we discuss the effect of seagulls (if there are any, $\delta(2-1)$ type terms) later and omit them for a while.

$$\{B(2)A(1)\}_T = \theta(t_2-t_1)B(2)A(1)+\theta(t_1-t_2)A(1)B(2)$$

If in particular we are interested in the forward scattering amplitude $T_{BA}(q)$ on a proton with a photon of momentum q we need the Fourier transform

$$-iT^F_{BA}(\nu, \vec{Q}) = \text{F.T.}[\theta(t_2-t_1)B(2)A(1)+\theta(t_1-t_2)A(1)B(2)] \tag{36.8}$$

(The superscript F indicates that the choice of sign of the imaginary part for negative frequencies is taken according to the convention of Feynman in his QED papers; there is a different choice called causal amplitude, T^C_{BA}, which is often very useful.)

To take the Fourier transform of the first term in (36.8) we have a product of $\theta(t_2-t_1)$ whose F.T. is $i/(\nu+i\varepsilon)$ and B(2)A(1) whose F.T. is $K_{BA}(\nu, \vec{Q})$ and hence we have the convolution of these. In the same way the second term is the convolution of $-i/(\nu-i\varepsilon)$ (the F.T. of $\theta(t_1-t_2)$ and $K_{AB}(-\nu, -\vec{Q})$.) Hence

$$T^F_{BA}(\nu, Q) = \int \left[\frac{1}{\nu'-\nu-i\varepsilon} K_{BA}(\nu', \vec{Q}) - \frac{1}{\nu'-\nu+i\varepsilon} K_{AB}(-\nu', -\vec{Q})\right] \frac{d\nu'}{2\pi} \tag{36.9}$$

Now use $\dfrac{1}{\nu'-\nu \mp i\varepsilon} = P\dfrac{1}{\nu'-\nu} \pm i\pi\delta(\nu'-\nu)$

$$T^F_{BA}(\nu, \vec{Q}) = \int \left(P\frac{1}{\nu'-\nu}\right) C_{BA}(\nu', \vec{Q}) \frac{d\nu'}{2\pi} + \frac{i}{2}\,\text{sgn}(\nu)\, C_{BA}(\nu, \vec{Q}) + \sigma_{BA} \tag{36.10}$$

$$\text{sgn}(\nu) = \begin{cases} +1 & \nu > 0 \\ -1 & \nu < 0 \end{cases}$$

where we have explicitly added possible seagull terms σ_{BA} which is simply an unknown (finite order) polynomial in ν, $\vec{Q}$. Thus the scattering amplitude is, except for a polynomial, given in terms of the commutator.

Negative ν, is of course, not defined by experiment. However it can be obtained from measurements of the reaction with antiparticles (using $\bar{A}$ for A) via the connection implied by (36.10) and the relation for commutators (resulting from the fact that $\bar{A}$ is the adjoint operator to A)

$$(C_{BA}(\nu, \vec{Q}))^* = C_{\overline{AB}}(\nu, \vec{Q}) = -C_{\overline{BA}}(-\nu, -\vec{Q})$$

so that (36.10) implies

$$T^F_{BA}(\nu, \vec{Q}) = T^F_{AB}(-\nu, -\vec{Q})$$

Lecture 37

Properties of operators, continued

Another more usual convention (causal amplitude) to define the negative frequency extension changes the sign of the imaginary part for negative frequencies

$$T^C_{BA}(\nu, \vec{Q}) = T^F_{BA}(\nu, \vec{Q}) + i\theta(-\nu)\, C_{BA}(\nu, Q) \quad . \tag{37.1}$$

Thus $(T^C_{BA}(\nu, Q))^* = T^C_{\overline{BA}}(-\nu, -\vec{Q})$

and also (37.1) becomes now

$$T^C_{BA}(\nu, \vec{Q}) = \int C_{BA}(\nu', \vec{Q}) \frac{d\nu'/2\pi}{\nu'-\nu-i\varepsilon} + \sigma_{BA} \tag{37.2}$$

The significance of T^C_{BA} (which is just as good a way of describing scattering as T^F) in coordinate space can now be seen. (T^C is given by an expression like (36.9) except that the sign of the $+i\varepsilon$ in the last term is reversed.) In (37.1) we have expressed things in terms of the commutator but the last term can be more simply expressed as the product operator, from (36.5) and (36.6) we have $\theta(-\nu)\, C_{BA}(\nu, \vec{Q}) = -K_{AB}(-\nu, -\vec{Q})$, the F.T. of A(1)B(2). Hence (37.1) says

$$-iT^C = \text{F.T. } \{B(2)A(1)\}_T - \text{F.T.} A(1)B(2)$$

for $t_2 > t_1$ we have $B(2)A(1) - A(1)B(2) = [B(2), A(1)]$ and for $t_2 < t_1$, we have $A(1)B(2) - A(1)B(2) = 0$

Hence the scattering amplitude(causal)is the Fourier transform of the retarded commutator + seagulls. By retarded commutator we mean the commutator for $t_2 < t_1$:

$$-iT^C_{BA}(q) = \text{F.T.}(\langle p|[B(2),A(1)]|p\rangle\, \theta(t_2-t_1)) + \text{seagulls} \tag{37.3}$$

from this, (37.2) is directly obvious.

(Remark. The $\theta(t_2-t_1)$ in 37.3 at first sight makes the result not relativistically invariant, until it is realized that the commutator factor is zero for spacelike regions so the plane $t_2 = t_1$ can be tilted arbitrarily as required by Lorentz transformation. This is true at least if the commutator is not too singular at equal times which is often true. Difficulties sometimes arise making T^C_{BA} defined here not relativistically invariant unless corresponding non-invariant terms (called Schwinger terms) are added into the "seagull" part of (37.3).)

Expression (37.3) serves as a general definition of the Chew amplitude, i.e. in space-time it is the retarded commutator, even when we do not have the diagonal elements, or lowest states (so the product operator is 0 for $\nu < 0$). The general definition of the Feynman amplitude is the time ordered operator (36.8). They differ by the product operator.

Note on various relations

Reality conditions

$$K_{BA}(2,1) = K_{\overline{AB}}(1,2)$$

$$C_{BA}(2,1)^{\dagger} = C_{\overline{AB}}(1,2) \tag{37.5}$$

$$\{K_{BA}(\nu, \vec{Q})\}^{*} = K_{\overline{AB}}(\nu, \vec{Q})$$

$$\{C_{BA}(\nu, \vec{Q})\}^{*} = C_{\overline{AB}}(\nu, \vec{Q}) = -C_{\overline{BA}}(-\nu, -\vec{Q}) \tag{37.6}$$

Crossing

$$\{T^C_{BA}(\nu, \vec{Q})\}^{*} = T^C_{\overline{BA}}(-\nu, -\vec{Q})$$

$$T^F_{BA}(\nu, \vec{Q}) = T^F_{AB}(-\nu, -\vec{Q}) \tag{37.7}$$

Imaginary Part

$$\{T^C_{\overline{BA}}(\nu, \vec{Q})\}^{*} - T^C_{AB}(\nu, \vec{Q}) = -iC_{AB}(\nu, \vec{Q}) \tag{37.8}$$

The σ_{BA} satisfy reality and crossing relations required to keep them valid for T. Namely

$$\sigma_{BA}^{*}(\nu, \vec{Q}) = \sigma_{\overline{BA}}(-\nu, -\vec{Q}) = \sigma_{\overline{AB}}(\nu, \vec{Q})$$

The commutator at equal times $t_2 = t_1$ can be obtained by integrating $C_{BA}(\nu, \vec{Q})$ with respect to ν for all ν, since $\int e^{i\nu(t_2-t_1)} d\nu/2\pi = \delta(t_2-t_1)$. Hence Gell Mann's equal time commutation relation

$$[J^a_o(t_1\, \overline{x}_2),\ J^b_\mu(t_1\overline{x}_1)] = \delta^3(\overline{x}_2-\overline{z}_1)\ J^{axb}_\mu(t,\overline{x}_1) \tag{37.9}$$

becomes upon Fourier transform for our diagonal element on the proton,

$$\int_{-\infty}^{\infty} C_{J_o^a J_\mu^b}(\nu, \vec{Q})\, d\nu/2\pi = \langle p|J_\mu^{axb}(0, 0)|p\rangle \tag{37.10}$$

a constant, independent of $\vec{Q}$. This is called a sum rule.

We now return to our study of the fact that in measuring $K_{BA}(\nu, Q)$ we are measuring the F.T. of the commutator of two currents. As examples of questions we shall ask are (1) What limitations on the F.T. result from the fact that the commutator vanishes outside the light cone? (2) From experimental facts about the behavior of K (e.g. Bjorken scaling) what do we learn about the character of the commutator?

It behooves us to study the general behavior of commutators, and we begin our study with the commutator of two scalar fields of mass m in a system without interaction:

$$C_m(2-1) = [\phi(2), \phi(1)] \tag{37.11}$$

We can express $\phi(1)$ in terms of creation and annihilation operators in the usual way

$$\phi(t,\bar{x}) = \sum_{\bar{k}} (2\omega_k)^{-1/2} \left(a_k e^{i\bar{k}\cdot\bar{x}-i\omega_k t} + a_k^* e^{-i\bar{k}\cdot\bar{x}+i\omega_k t} \right) \tag{37.12}$$

Here $\omega_k = \sqrt{k^2+m^2}$ is the correct frequency to describe the operator's development in time for there is no interaction so the energy is that of a free particle. The a_k's commute with each other, and the a_k^*'s, only a and a* do not commute if they belong to the same $\bar{k}$:

$$[a_{\bar{k}}, a_{\bar{k}'}^*] = \delta_{\bar{k}\bar{k}'} \tag{37.13}$$

We work out (37.11) immediately by forming the commutator, using (37.13), of an expression like (37.12). Dropping terms which obviously commute we are left with:

$$C(2-1) = C(t,\bar{x}) = \sum_{\bar{k}} \frac{1}{2\omega_k} \left(e^{-i\omega_k t} e^{i\bar{k}\cdot\bar{x}} - e^{i\omega_k t} e^{-i\bar{k}\cdot\bar{x}} \right)$$

$$= \int \frac{d^3k}{(2\pi)^3 2\omega_k} \left(e^{-i\omega_k t} e^{i\bar{k}\cdot\bar{x}} - e^{i\omega_k t} e^{-\bar{k}\cdot\bar{x}} \right) \tag{37.14}$$

The actual integral can now be done - it involves Bessel functions.

Here we will just do the special case m = 0, and say what the case $m^2 \neq 0$

gives, leaving details to the student.

In (37.14) put $\overline{k}\cdot\overline{x} = kR\cos\theta$, $\omega_k = k$ for $m = 0$ to get

$$C(t,\overline{x}) = \int \frac{2\pi k^2 dk\, d\cos\theta}{(2\pi)^3 2k}\, e^{-ikt} e^{ikR\cos\theta} + C.C.$$

$$= \frac{-i}{2\pi}\, \frac{1}{2R}\, (\delta(t-R)-\delta(t+R)) = \frac{-i}{2\pi}\, \delta(t^2-R^2)\ \mathrm{sgn}(t)$$

singular δ function on the light cone.

For finite mass we obtain the same singularity on the light cone, zero outside it of course, and a Bessel function inside:

$$C_m(t,x) = i\,\mathrm{sgn}\ t\ [-\frac{\delta(s^2)}{4\pi} + \frac{1}{8\pi s}\ J_1(ms)\theta(s^2)]$$

Thus we see that the commutator for free particles is zero outside the light cone and singular (like δ function) on the light cone.

We have discussed the singularities of two fields. Also instructive is to discuss the singularities from two currents. One such current might be in a free field theory $J_\mu(1) = \overline{\phi}(1)\ (\overleftarrow{\partial}_\mu + \overrightarrow{\partial}_\mu)\phi(1)$ or $\psi(1)\gamma_\mu\psi(1)$ for spinor fields. As an example we will leave out these gradients, etc. and find the commutator for "currents"that are simply squares of a scalar field

$$K = [J(2),\ J(1)] = [\phi(2)\phi(2),\ \phi(1)\phi(1)] = 2\phi(2)[\phi(2),\ \phi(1)]\phi(1) +$$

$$+\ 2\phi(1)[\phi(2),\ \phi(1)]\phi(2)$$

$$= 2\phi(2)\ C_m(2,1)\phi(1) + 2\phi(1)\ C_m(2,1)\phi(2)$$

where $C_m(2,1)$ is the free field commutator we worked out before. Clearly K has the same singularity sgn $(t_2-t_1)\delta(s_{12}^2)$ as noted before for C, but this time multiplied by an operator function of two positions F(2,1) called often a bilocal operator. This operator is needed, of course, only on the light cone. Its matrix elements give functions of $x_{2\mu}-x_{1\mu}$ (for example of t_2-t_1 for a diagonal element in a system of a particle at rest). So in general the singularities along the cone are modulated by a function of the distance from the origin of the cone.

For interacting hadrons the commutator of currents is expected to be zero outside the light cone also, and non-zero only inside. An interesting question is how it makes the transition as we cross the light cone. Some sort of jump is expected presumably. Is it in value or only in slope? Or possibly it has δ function as the free particle case does. Such a question is a fundamental one only answerable by experiment.

We have the Fourier transform experimentally in $K_{\mu\nu}(\nu, \vec{Q})$. A singular behavior corresponds to some sort of high ν, high Q limit; so it is the behavior at large ν, Q that gives us the answer to the question. But in this region experiments indicate K(ν, Q) satisfies Bjorken scaling - that is, it is only a function of $\xi = -q^2/2M\nu$.

The most straightforward way to analyze this is to take the inverse transform of K, using the Bjorken limit and seeing what singularities it gives - then concluding the character of the singularities, because they depend only on the high ν, Q limit are the same for the complete function as they are for the Bjorken limit. We will not try to be rigorous, because for one thing we really do not know, experimentally, K for $q^2 > 0$, nevertheless we can see what happens.

Let Q be along the z axis. $-q^2 = Q_z^2 - \nu^2 = (Q_z+\nu)(Q_z-\nu) = 2M\nu\xi$, hence for large ν, Q_z is nearly ν in fact $Q_z = \nu + M\xi$. The Fourier inverse of a function of ξ only would look like

$$\frac{1}{(2\pi)^2}\int e^{i(\nu t - Q_z z)} f(\xi)\,d\nu\,dQ_3 = \frac{1}{(2\pi)^2}\int e^{iQ_z(t-z)}\,e^{-iM\xi t}\,f(\xi)\,d\xi\,dQ_z$$

$$= s(t)\,\delta(t-z)$$

where s(t) is the F.T. of the structure function f(ξ).

To put it in another slightly more rigorous way we note

$$f(-q^2/2M\nu) = 2M\nu\int \delta(q^2+2M\nu\beta)\,f(\beta)\,d\beta$$

Now omit for a moment the 2Mν (suppose we took the transform of W_2) and note for large ν, q^2 that to a good enough approximation the $\delta(q^2+2M\nu\beta)$ can be replaced by $\delta(q^2+2M\nu\beta+M^2\beta^2)$. So we get for large enough ν, $(q^2+2M\nu\beta+M^2\beta^2 = (q+\beta P)^2)$

$$W_2 = \int \delta((q+\beta P)^2)\,f(\beta)\,d\beta$$

Now we can insert, if we wish, a sgn(ν) to keep the asymmetry as is required for a F.T. of a commutator, and for large ν this is just sgn(ν+βM) (if the proton is at rest). Hence asymptotically

$$W_2 = \int \text{sgn}(\nu+\beta P_o)\delta((q+\beta P)^2)f(\beta)d\beta$$

Now the F.T. is easy for we know that

$$\text{F.T. sgn}(t)\delta(s^2) = \text{sgn}(\nu)\delta(q^2) \qquad s^2=t^2-R^2$$

so that for a four vector a

$$\text{F.T. } e^{-ia\cdot x}\text{sgn}(t)\delta(s^2) = \text{sgn}(\nu+a_o)\delta((q+a)^2)$$

hence

$$\text{F.T.}W_2 = \int f(\beta)d\beta\ e^{-i\beta P\cdot x}\text{sgn}(t)\delta(t^2-R^2)$$

noting $P\cdot x = Mt$ we have our previous result - that the significance of Bjorken scaling is that the singularity of the current operators has a δ function-like singularity on the light cone. To be more precise we will have to include all the $P_\mu P_\nu$ factors, etc., and define everything precisely - there will then be gradients of δ-functions involved. The best way to say it in a general way is that scaling shows that the singularities on the light cone of the current commutators are just of the same severity as they would be for free particle fields.

This is, of course, what we expect from the parton interpretation for there we assumed in the final states x(which become the intermediate states of the commutator) the partons can be considered as free. Thus there is no surprise in the conclusion, it is only that we wish to state what we have discussed (scaling) in an abstract and general way (as current commutator singularities) with minimum references to a model. Although the statement still seems to refer to free particle fields, that is only a shorthand to writing δ and δ' functions on the light cone.

Each general property of partons we assumed said something more explicit about the character of the singularity. For example, to say charged partons are spin 1/2 is to say the singularities are like those δ or δ' functions characteristic of free Dirac field commutators. In addition, vector and axial vector results are related. If, for example, one adds that partons are quarks certain numerical relations are implied, as we have seen, between the singular parts of the commutators of various kinds of currents. (That is to say for example, our relations among f^{ep} and $f_1^{\nu p}$, $f_3^{\mu p}$ etc. implied by expressing everything in terms of the six functions u, $\bar{u}$, d, $\bar{d}$, s, $\bar{s}$.)

Quarks, as free particles, have not been found. There are many questions as to whether the detailed views of the parton model are correct for quarks. In particular, the question whether a single parton quark moving out in recoil need or need not take its non-integral quantum number with it.

In addition it is often very useful to see, when results have been obtained from a model, just how much depends on the model and whether in fact the results cannot be stated as a general mathematical principle without recourse to a specific model. In that way, if it proves later that too many details of the model are faulty, we can still begin again having learned some general properties of hadrons without being committed to all the others.

Therefore it is interesting to note that the parton model is equivalent to the general statement (and one much closer to direct experimental results) that commutators have δ-like light-cone singularities.

Light Cone Algebra

Lecture 38

Light Cone Algebra

An interesting question answered by Fritzsch and Gell-Mann (1971 Coral Gables conference) is how the scaling results of the "partons as quarks" can be stated in a way which - at the end - doesn't involve quark wave functions or operators at all.

We wish to state that the light cone singularities are like those for free quark (spin 1/2, SU_3 triplet) commutators. We mean, of course, the largest δ' or δ type of singularity, as we say the "leading" (in high frequency) singularity. First we see what the singularities are like if currents were represented by quark fields $\psi(x)$, a Dirac spinor carrying SU_3 indices on which 3 x 3 matrices λ can act.

The commutator of two spinor fields of mass m is easily worked out (as we did for Bose fields) it is

$$[\psi(2), \bar{\psi}(1)] = (i \not{\nabla}_2 + m)\, C_m(2, 1)$$

(Just as the propagator is $(\not{p} + m)/(p^2 - m^2)$ instead of $1/(p^2 - m^2)$ so this $\not{p} + m$ comes likewise into the commutator (the real part of the propagator). Since we are looking for the leading singularity near the light cone C_m is

$\delta(s^2)$ and the mass term is smaller than the large gradient, thus

$$[\psi(2), \bar{\psi}(1)] \hat{=} \frac{i}{4\pi} \not{\nabla}_2 \, \mathrm{sgn}(t_2 - t_1)\delta(s_{21}{}^2) \tag{38.1}$$

where $s_{21}{}^2$ is the square interval from 1 to 2 and $\hat{=}$ means "the leading singularities near the light cone are equal."

It is easily verified that the results we obtain here for the commutators of two currents each a bilinear form in $\bar{\psi}$, are exactly the same whether we suppose a) the fields obey the usual anticommutation relations at equal times $\psi(\bar{x}_2)\bar{\psi}(\bar{x}_1) + \bar{\psi}(\bar{x}_1)\psi(\bar{x}_2) = \delta^3(\bar{x}_1 - \bar{x}_2)$ and $\psi(\bar{x}_1)$, $\psi(\bar{x}_2)$ anticommute as is appropriate to Fermi particles; or we suppose b) the spin 1/2 fields obey commutation relations $\psi(\bar{x}_2)\bar{\psi}(\bar{x}_1) - \bar{\psi}(x_1)\psi(x_2) = \delta^3(\bar{x}_2 - \bar{x}_1)$ and $\psi(\bar{x}_1)\psi(\bar{x}_2)$ commute as is appropriate to Bose particles.

This is very interesting because it says the "Bose quark" model which is appropriate at low energies is in no fundamental way in contradiction to "partons as quarks" at higher energy. We can call this parton model "partons as Bose quarks."

Next a current of SU_3 type a (described by λ^a) is

$$J_\mu^a(1) = \bar{\psi}(1)\lambda^a\gamma_\mu\psi(1) \tag{38.2}$$

where for example for electric current λ^a is diagonal say $\lambda^\gamma = \mathrm{diag.}(2/3, -1/3, -1/3)$. For axial currents change γ_μ to $\gamma_\mu\gamma_5$.

Now when we commute two currents, we find a simple but tediously complicated result (see reference for details). To illustrate the idea, which is all we intend to do here, we take the case of two electric currents, and also drop several terms which would vanish if we took spin averaged matrix elements, then all that survives is

$$[J_\mu^\gamma(2), J_\nu^\gamma(1)] \hat{=} \frac{1}{4\pi}\partial_\rho \left(\varepsilon(t_2-t_1)\delta(s_{12}^2)\right)\frac{1}{4}T_r(\gamma_\mu\gamma_\rho\gamma_\nu\gamma_\sigma)[\mathcal{F}_\sigma^\delta(2,1) - \mathcal{F}_\sigma^\delta(1,2)] \tag{38.3}$$

(plus other terms dropped)

where

$$\mathcal{F}_\sigma^\delta(2,1) = \bar{\psi}(2)\gamma_\sigma\lambda^\delta\psi(1) \tag{38.4}$$

(2,1 on mutual light cone) and λ^δ is diag.(4/9, 1/9, 1/9) $= \lambda^\gamma\lambda^\gamma$ in our case.

Now we still see the quark fields in (38.4), but the idea now is to disregard equation (38.4) and to suggest that equation (38.3) (its generalization to arbitrary currents and including omitted terms of course) are generally valid.

They give the light cone singularities and define, on the light cone at least a set of new operators $\hat{\mathcal{F}}_\sigma(2\text{-}1)$. It is the matrix elements of these operators which give us our structure functions.

No direct implication is made that they can be expressed as (38.4). Equation (38.4) as well as (38.1), (38.2) are just scaffolding to arrive at (38.3) and are henceforth to be forgotten. In (38.3) no explicit quark operators are seen. Instead only some new bilocal operators are defined (meaning depending on two points). They are defined only on the light cone by the very equations (38.3)

[This system would be a true "algebra" if the properties of the $\hat{\mathcal{F}}$ could now be defined independently. For example, an ideal situation (in fact a complete theory of the hadrons) would result if equations giving the commutators of such $\hat{\mathcal{F}}$'s could be given in terms of J's and $\hat{\mathcal{F}}$'s themselves. This does not seem possible. A little can be done, however. With the form (38.4) the commutators of two $\hat{\mathcal{F}}$'s which have their variables on the same light ray (e.g. $[\hat{\mathcal{F}}(3,4), \hat{\mathcal{F}}(1,2)]$ where 3,4,1,2 all lie on a single generator of a light cone) again can be expressed as an $\hat{\mathcal{F}}$. This relation has been also hypothesized, but it is not much. It gives predictions for certain two-current inclusive reactions like $e + p \to e + \mu^+ + \mu^-$ + any hadrons. They are, of course, just those expected from the parton model in the same conditions. We discuss them later.]

In order to use equation (38.3) we must take the Fourier transform, and for that we can use the result (very similar to those we have already derived, we leave details to you) if $q = (\nu, 0, 0, \nu + M\xi)$ <u>for large ν</u>.

$$\int d^4x\, e^{iq\cdot x}\delta'(x^2)F(x_o, x_1, x_2, x_3)$$

$$\pi\int \frac{dt}{|t|}\, e^{-iM\xi t}F(t, 0, 0, t)$$

that is, it involves only the integral along the ray $t = z$ common to the plane $t = z$(from $e^{i\nu(t-z)}$) and the light cone.

Naturally every result of this theory is also a result of the "parton as quark" theory for the latter is a model of the former; but not every parton result can be derived from (38.3). It is possible, therefore, that many of them are wrong and only (38.3) and not the complete field model may survive.

Therefore it is interesting to compare the theories for various types of predictions made by the parton as quarks theory. They seem to be of three classes.

A. Scaling, relations among scaling functions, sum rules.

B. Special arguments about the structure functions derived by arguments about hadron collisions and other arguments; like dx/x behavior, relation of form factor power $1/Q^{\gamma}$ to x near 1, structure function $(1-x)^{\gamma}$ etc.

C. Applications to experiments of Drell type $e^+ + e^- \to$ hadrons or $p + p \to \mu^+\mu^-$ + any hadrons, etc.

Class A are exactly those derived from light-cone algebra. Class B are not obtained from light cone algebra by default. That is, they simply amount to a discussion of how the matrix element of $\hat{\mathfrak{F}}(2,1)$ might behave. It is not a specific assumption of the model but an attempt to go further that leads to Class B results - a serious attempt to discuss the matrix elements of the light cone algebra would lead to results in this class.

Class C are very interesting, as they seem beyond (38.3) and require some extension, even though they appear evident from the parton view. It is therefore here, testing these that a real choice can be made as to whether the extensions of the parton model beyond the expectations of (38.3) are really sound.

The reason the current commutator on the light cone is not sufficient for reactions like $p + p \to \mu^+ + \mu^- + X$ is that we do need matrix elements like

$\langle pp|JJ|pp\rangle$

but this time as we take the limit as the momentum of the current operator increases we are also changing the state pp; the relative momentum of the two protons must also increase. This is an awkward limit for a theory of operators.

It is a very interesting question. Is there some general abstract way without quark fields (partons) to describe all these Class C parton predictions also? Or are they perhaps wrong, unreliable extensions of the idea?

And what becomes of the question of how the partons come apart in the proton without exhibiting quark quantum numbers among the final states? Does current algebra help us to solve it? Perhaps, it seems to be translated into "are there any representations possible of the algebra (38.3) which do not imply quark quantum numbers among the localized states? It may be more tractable in this mathematical form than in physical arguments.

Properties of Commutators in Momentum Space

Lecture 39

Properties of Commutators in Momentum Space

We now come to discuss what general properties a commutator has to have in momentum space, such that its Fourier transform will be zero outside the light cone of configuration space.

This is clearly a promising line to follow to supplement physical intuition on the properties of $W_{\mu\nu}$. At present not a great deal can be said, but for your interest in your future research what has been done here may prove fruitful.

First we can make some obvious remarks. If we multiply a commutator (in space-time) by a function G which is 1 inside and zero outside a light cone we recover the commutator (except, as is the case, that it is singular right on the light cone where our function G is poorly defined). Since the F.T. of G is $16\pi \text{P.V.}(1/q^2)^2$ (P.V. = principal value) we have the convolution theorem

$$C(q) = \int C(u)\text{P.V.}\frac{16\pi}{(q^2-u^2)^2}\frac{d^4u}{(2\pi)^4} \tag{39.1}$$

(plus pieces from the light cone). (C(q) is the F.T. of a commutator.)

In a similar vein, since the F.T. of 0 0 is $16\pi i\,\text{sgn}(q_o)\delta'(q^2)$

we see that C must have the form

$$C(q) = \int X(u)\,\mathrm{sgn}(q_o-u_o)\,\delta'((q-u)^2)\,\frac{d^4u}{(2\pi)^4}\,(16\pi i) \tag{39.2}$$

plus light cone pieces. Of course we can say also in (39.1) that the C(u) of the integrand is any function F(u) at all, and the C that comes out on the left will be the F.T. of a function zero outside the light cone. I have not been able to make much use of these observations.

In using the simple views above we must be careful of one point. The function we multiplied C(x,t) by, G(x,t), was defined as 1 inside the light cone and 0 outside; what is it exactly on the light cone? It is indefinite there, ordinarily that is such a small region that it makes no difference in the integral over space time of C(x,t) G(x,t) but in fact it makes a great uncertainty because C(x,t)has a δ function singularity just where G(x,t) is poorly defined. We can correctly straighten out our formulas for this effect in the following way. We call $C_a(q)$ the asymptotic C(q) giving just the light-cone δ singularities. Than $C(q)-C_a(q)$ has no light cone singularity and so equation (39.1) for example, holds if C is replaced by $C-C_a$ on both sides. One can then simplify or rearrange the equation to obtain one like (39.1) but with some additional terms related to the light cone behavior.

A far more subtle and useful observation was made by Dyson. He noticed, in many problems we also know C(q) = 0 for certain regions of q space (where no intermediate states may be available). For example, for $q_o = \nu > 0$ the lowest state available for the final state x is the proton itself, moving with momentum Q (space part of q_μ) hence with energy $\sqrt{M^2+Q^2}$ hence

$$\nu > -M + \sqrt{M^2+Q^2}$$

and likewise

$$\nu < +M - \sqrt{M^2+Q^2}\;. \tag{39.3}$$

Dyson has proved that the necessary and sufficient condition for C(q) to vanish in region $S = S_1(q) < q_o < S_2(q)$ and to have a Fourier transform vanish outside the light cone is that C(q) can be written as

$$C(q) = \int d^4u \int_0^\infty ds^2\;\mathrm{sgn}(q_o-u_o)\,\delta((q-u)^2-s^2)\,\Phi(u,s^2) \tag{39.4}$$

(Dyson Representation)

where Φ vanishes outside a region R, but is otherwise arbitrary. The region R is such that the $(q-u)^2-s^2$ hyperboloid does not penetrate region S.

That it is a sufficient condition is easy to see (the great difficulty of the proof is to show it is necessary). We have already seen that the free particle commutator $C_m(x,t)$ is zero outside the light cone, so the F.T. of $(q^2-m^2)\,\mathrm{sgn}(q_o)$ is zero outside the light cone. But if $C_m(x,t)$ is multiplied by anything, in particular by $e^{iu\cdot x}$ (or any superposition over u) it is still zero outside the light cone, hence F.T. $e^{iu\cdot x}C_m(x,t) = \mathrm{sgn}(q_o-u_o)\delta((q-u)^2-m^2)$. Hence superposing with weight $\Phi(u,s^2)$ various cases of u and $m^2 = s^2$ we get (39.4)

It is easy to see, by drawing hyperbolas that for our case the region R of integration where $\Phi(u,s^2)$ does not vanish is as follows

For $s^2 > M^2$ inside two cones $|u_o|<|\bar{u}|$

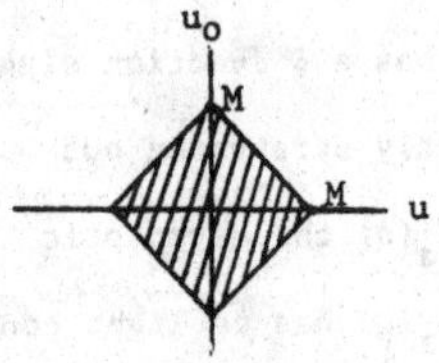

For $s^2 < M^2$ inside region bounded by $M - |u_o| = \sqrt{\bar{u}^2 + (M-s)^2}$

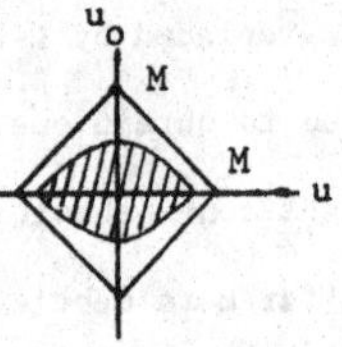

Incidentally, the scattering amplitude that corresponds to this commutator (using 37.2) becomes

$$T^c(q) = \frac{1}{2\pi}\int_R d^4u \int ds^2 \frac{\Phi\,(u,s^2)}{(q-u)^2-s^2+i\varepsilon(q_o-u_o)} + \sigma(q) \qquad (39.5)$$

where σ (from seagulls) is a polynomial in q_μ (for the amplitude T^F replace $i\varepsilon(q_o-u_o)$ by $i\varepsilon$).

One of the difficulties in using the Dyson representation is that the function Φ is not unique, many can give the same C(q). Another very similar representation has been derived (I suspect not as rigorously as Dyson's) by Deser, Gilbert and Sudarshan especially for a problem like ours where C is a function only of the two invariants q^2, ν, it is

$$C(q^2,\nu) = 2M\nu \int_0^\infty d\sigma \int_{-1}^{+1} d\beta\; H(\sigma,\beta)\,\mathrm{sgn}(q_o+\beta M)\,\delta((q+\beta p)^2-\sigma) \qquad (39.6)$$

This is, of course, just Dyson's representation if the four vector u_μ can be assumed to have only a time component, that is, only a component in the p_μ direction.

There is again the expectation that $H(\sigma,\beta)$ is zero outside a more limited range if $\sigma < M^2$. In fact if $\sigma < M^2$, β only runs from $-\sqrt{\sigma}/M$ to $+\sqrt{\sigma}/M$. but we leave our expressions intact, just remembering $H(\sigma,\beta) = 0$ for $|\beta| > \sqrt{\sigma}/M$ in this region of σ.

It is clear that if C is zero on the light cone various gradients are also, so several possibilities exist with different integral powers of ν (from zero up to any finite value) with corresponding H functions; but we have taken the one appropriate to $2MW_1$. Equivalently we can write our form with a new definition for σ.

$$2MW_1(q^2,\nu) = 2\pi\nu \int_0^\infty d\sigma \int_{-1}^{+1} d\beta h(\sigma,\beta)\,\mathrm{sgn}(\nu+M\beta)\,\delta(q^2+2\beta M\nu-\sigma) \tag{39.7}$$

In addition, for our case, the asymmetry in time for the commutator becomes the property that

$$h(\sigma,-\beta) = -h(\sigma,\beta) \tag{39.8}$$

This representation is very nice, and h is probably uniquely determined by W_1. The weakness, however, is that no physical interpretation or expression in terms of matrix elements is given for $h(\sigma,\beta)$. Therefore we cannot use any physical intuition in guessing how h should behave (other than using our knowledge of how W behaves and working backward). That is to say we cannot at any stage say - such a function for $h(\sigma,\beta)$ is too "crazy" physically - or it ought to behave so and so in this region, etc. As we shall see, this is a serious weakness in this case.

We now turn to see how $h(\sigma,\beta)$ must behave in order to produce a function $2MW_1$ behaving as we expect (see Lecture 31)for large ν (we take $\nu > 0$ throughout).

<u>Region I</u>

First of all we have the scaling limit $\nu \to \infty$, $-q^2 \to \infty$, $-q^2/2M\nu = x$ finite. Here $2MW_1 = f(x)$ a function of x only. Eq.(39.7) gives ($\mathrm{sgn}(\nu+\beta M) = 1$ for large ν, since $\nu > M$).

$$2MW_1 = 2M\nu \int_0^\infty \int_{-1}^{+1} d\beta \; h(\sigma,\beta)\delta(-2M\nu x + 2M\nu\beta - \sigma)d\sigma$$

Now if we assume $h(\sigma,\beta)$ falls away rapidly enough with σ only finite σ are involved here, so as $\nu \to \infty$ the σ can be dropped in the δ function and we have $h(\sigma,x)d\sigma$ a function of x as required

$$f(x) = \int_0^\infty h(\sigma,x)d\sigma \qquad (39.9)$$

I believe it was this argument with the Dyson representation which either led Bjorken to his scaling hypothesis, or helped to confirm his suspicions of its truth.

$q^2 > 0$ scaling

We have here a bonus. We can now determine how the function $2MW_1(q^2,\nu)$ should go in the positive q^2 scaling region. Letting $\nu \to \infty$, $q^2 \to \infty$ such that $+q^2/2M\nu = x'$ is finite we have $2MW_1 = \int_0^\infty h(\sigma, -x')d\sigma$ which again scales; but even more, from the symmetry of h it scales to the same function $f^{ep}(x')$

$$2MW_1 = -f^{ep}(x') \qquad q^2 > 0$$

How can we see this remarkable result from the parton model? Can we equally well derive the function for $q^2 > 0$ by using physical arguments? We will explain qualitatively how it comes about.

For positive q^2 we must be concerned in the commutator (36.1) with the subtracted vacuum piece. In the vacuum we can make pairs, and at high q^2 just a pair of partons

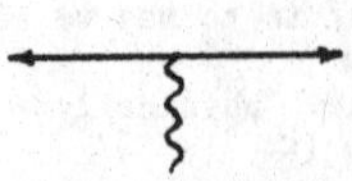

We are concerned with how this pair production probability is modified by the presence of a proton

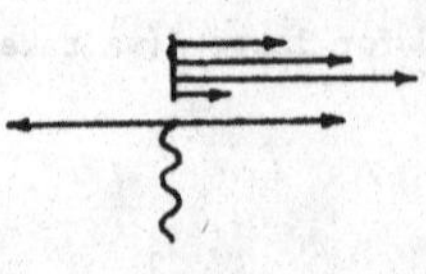

At first one might think the modification must come through interaction of

the newly made partons with those in the proton - in interaction about which we know little except that it is finite and is not involved in our derivation of $f(x)$ for negative q^2. But at high energy a more important effect is the exclusion principle - partons cannot be made if they are already present in the proton (if Fermi statistics is assumed). Thus in our diagram the probability to produce a $\bar{u}$ to the left and u to the right at x' is 4/9 units (for the charge is 2/3). However, it is altered at those x by the chance, $u(x)$, that a u parton is present in the proton. Thus we have a contribution proportional to $(4/9u(x')$; if the $\bar{u}$ parton is moving forward it is $(4/9)\bar{u}(x')$. If a d parton pair is produced the probability is 1/9, modified by the chance of finding a d parton already in the proton, the contribution is $(1/9)d(x')$, etc.) Thus the entire contribution is $4/9[u(x')+\bar{u}(x')]+ 1/9[d(x')+\bar{d}(x')] + 1/9[s(x')+\bar{s}(x')]$ or $f^{ep}(x')$ as required.

Bose or Fermi quarks

In making the analysis for q^2 positive we assumed the quarks obeyed Fermi statistics. If Bose statistics are used in the formal expression of the completed sum over states (36.1) we have some changes. First there is a change in sign of the Fermi exclusion effect turning into a Bose tendency to increase emission because a particle is already present. Then there is another change in sign in the closed-loop diagram $\langle 0|JJ|0\rangle$ when changing from Fermi to Bose statistics [see, for example, R.P. Feynman, Phys. Rev. 76 756 (1949)].

On the other hand with the Bose statistics sign the imaginary part of $\langle 0|JJ|0\rangle$ is negative and it cannot be written as a sum of positive probabilities $\sum_x |\langle 0|J|x\rangle|^2$, but is minus that sum. This circumstance is the basis of the usual proofs of the relation of spin and statistics; spin one-half particles cannot be consistently interpreted as Bose particles. The most straightforward interpretation of Bose quarks would mean the "Drell Constant" defined for the $e^+e^- \rightarrow$ any hadrons cross section would have the impossible value -2/3. Of course a completely naive interpretation of quarks as Bose particles would also lead to the wrong result that the wave function would be symmetrical under the interchange of two protons. If quarks obey para-statistics both of these problems would be removed.

Region II

We continue our discussion of equation (39.7) now turning to the region $\nu \to \infty$, $-q^2$ finite. Here we expect $2MW_1$ to go as ν times a function of q^2 which we have written as $2MW_1 = 2M\nu\, g(-q^2)/(-q^2)$. In this region our equation (39.7) reads

$$2MW_1 = 2M\nu \int_0^\infty d\sigma \int_{-1}^{+1} d\beta\, h(\sigma,\beta)\,\delta(q^2+2M\nu\beta-\sigma)$$

at first sight, for $\nu \to \infty$, we could forget $-q^2-\sigma$ in the δ and obtain $2MW_1 = \int_0^\infty h(\sigma,o)\,d\sigma$, a constant independent of q^2 and evidently incorrect. However, the "constant" would, by (39.9) be $f(o)$ which we know is infinite, as for small x, $f(x) \approx a/x$. This suggests that we assume for small β that $h(\sigma,\beta)$ is singular.

$$h(\sigma,\beta) = k(\sigma)/\beta \tag{39.10}$$

for small β. Substituting in the integral above gives

$$2MW_1 = 2M\nu \int_0^\infty \frac{k(\sigma)}{\sigma-q^2}\, d\sigma$$

Thus

$$\frac{g(-q^2)}{-q^2} = \int_0^\infty \frac{k(\sigma)}{\sigma-q^2}\, d\sigma$$

gives the $-q^2$ dependence of the total virtual photon cross section. The limit of $g(-q^2)$ as $-q^2$ gets large approaches $\int_0^\infty k(\sigma)\,d\sigma$.

Lecture 40

Region III

Finally we turn to region III: $\nu \to \infty$, large $-q^2$ but $M_x{}^2-M^2 = 2M\nu - (-q^2)$ kept finite. There we expect $2MW_1 = (-q^2)^{-\gamma} h(M_x{}^2-M^2) 2M\nu$ where h is a function of $M_x{}^2-M^2$ only. The $(-q^2)^{-\gamma}$ is typified by the elastic scattering $(M_x{}^2-M^2 = 0)$ and γ is perhaps 4 or 5. For x near 1, $f(x) \frown A(1-x)^{\gamma-1}$ and for large $M_x{}^2-M^2$, $h(M_x{}^2-M^2)$ goes as $A(M_x{}^2-M^2)^{\gamma-1}$.

In this region our equation (39.7) reads

$$2MW_1 = 2M\nu \int_0^\infty d\sigma \int_{-1}^{+1} d\beta\ \delta(M_x^2 - M^2 + 2M\nu(\beta-1) - \sigma) h(\sigma,\beta) \tag{40.1}$$

Again a "natural" guess (neglect $M_x^2 - M^2 - \sigma$ relative to $2M\nu(\beta-1)$ in the δ function) leads to the wrong answer: $2MW_1 = \int_0^\infty d\sigma\ h(\sigma,1)$, a constant independent of $M_x^2 - M^2$. But again we can fix it up by a more complicated guess for the behavior of $h(\sigma,\beta)$ near $\beta = 1$. Because, from (39.9) $\int_0^\infty d\sigma\ h(\sigma,x)$ is zero as $x \to 1$ (like $A(1-x)^{\gamma-1}$). Therefore one way to get the answer would be to assume that $h(\sigma,\beta)$ goes to zero near $\beta = 1$. For example suppose

$$h(\sigma,\beta) = D(\sigma)\ (1-\beta)^{\gamma-1} \qquad \text{as } \beta \to 1$$

then $\int_0^\infty D(\sigma) d\sigma = A$. (If we assume γ depends on σ we can be more general and get nearly the same final result except with some slow logarithmic dependences, which might in fact be there in this region. This is not known very well experimentally). Substituting into (40.1) and integrating (the δ function says $1-\beta = (M_x^2 - M^2 - \sigma)/2M\nu$) we get

$$2MW_1 = (2M\nu)^{-\gamma+1} \int_0^{M_x^2 - M^2} (M_x^2 - M^2 - \sigma)^{\gamma-1} D(\sigma) d\sigma \tag{40.2}$$

hence of the correct form if (call $M_x^2 - M^2 = \lambda$)

$$h(\lambda) = \int_0^\lambda (\lambda-\sigma)^{\gamma-1} D(\sigma) d\sigma \tag{40.3}$$

This appears at first satisfactory; h will have the right asymptotic behavior $A\lambda^{\gamma-1}$ for large λ if $D(\sigma)$ falls off fast enough with σ, and $\int_0^\infty D(\sigma) d\sigma = A$.

However, we will again be fooled in our qualitative expectations for $h(\lambda)$ from this equation if we suppose that $D(\sigma)$ is a simple function. The operation of integration in (40.3) is a very powerful smoothing operation; for no simple $D(\sigma)$ would we be lead to guess what we know is true - that $h(\lambda)$ shows a series of peaks and valleys for small λ, the resonance peaks. It is not that (40.3) says resonance peaks are impossible - it is that (40.3) plus a guess that $D(\sigma)$ is not too complicated (i.e. does not vary in a rapid and peculiar way) fails to do so. To see this take the case $\gamma = 4$ and note that (40.3) can be solved as

$$D(\lambda) = \frac{1}{6} \frac{d^4}{d\lambda^4} h(\lambda) \tag{40.4}$$

The fourth derivative of a resonance oscillates in a very special manner. (For a much more intelligent way to deal with the resonance region see Cornwall, Corrigan and Norton, Phys. Rev. D3 536, (1971) and the next section.)

To summarize this investigation, the Deser, Gilbert, Sudarshan representation is useful as a formal tool to get dispersion relations, etc. It is, as of course it ought to be, capable of representing what we know of the experimental data with some function $h(\sigma,\beta)$. But it is disappointing in that its predictive utility - by guessing "reasonable" behavior for $h(\sigma,\beta)$ - is not very great first because we have no physical picture of what the function is, and second it seems like a function that behaves in a way as complicated as (if not more complicated than) the data itself.

It seems that $h(\sigma,\beta)$ is not, in fact, a function that has much direct physical significance and is purely an artificial expression in a certain mathematical form of the fact that the commutator vanishes outside the light cone.

It might pay to try to find another representation of this fact in which the kernel functions may have some direct physical representation.

Scattering in the Deser, Gilbert, Sudarshan representation

Given the commutator in any form we can find the scattering amplitude $T(q^2,\nu)$ corresponding to it by using the dispersion relation (37.2). It is easy to see by direct substitution of (39.7) that the scattering which has the right spin dependence (i.e. $\delta_{\mu\nu} q^2 - q_\mu q_\nu$ part) to correspond to $2MW_1$ (call it $2MT_1$) is

$$2MT_1 = -2M\nu \int_0^\infty d\sigma \int_{-1}^{+1} d\beta\, h(\sigma,\beta)(q^2+2M\nu-\sigma+i\varepsilon\ \mathrm{sgn}(\nu+M\beta))^{-1} + \text{seagulls} \tag{40.5}$$

$$2MW_1 = 2M\nu \int_0^\infty d\sigma \int_{-1}^{+1} d\beta\, h(\sigma,\beta)\,\mathrm{sgn}(\nu+M\beta)\,\delta(q^2+2M\nu-\sigma) \tag{40.6}$$

This is the causal amplitude convention, in the Feynman convention the $i\varepsilon\ \mathrm{sgn}(\nu+M\beta)$ is replaced simply by $i\varepsilon$. The "seagulls" are a polynomial in q^2,ν. This can be readily verified (simply notice that $\mathrm{sgn}(\nu+M\beta)\delta(q^2+2M\nu\beta-\sigma) = \mathrm{sgn}(\nu+M\beta)\,\delta((\nu+M\beta)^2-E^2) = (\delta(\nu+M\beta-E)-\delta(\nu+M\beta+E))/2E$ where we have written $E = +(\sigma+M^2\beta^2+Q^2)^{1/2}$).

Note 1: Because we are expressing T_1 as an integral with a ν factor times the integral we have really been working with W/ν and T/ν; but there is a pole in ν($W_1 \to 0$ as $\nu \to 0$ but T_1 does not). To avoid this use in our dispersion relation W_1/ν to get $(T_1(q^2,\nu)-T_1(q^2,0))/\nu$.

Note 2: It is evident from (40.5) and (40.6) that although derived from the dispersion relation in ν for fixed Q (space part of q) they also satisfy a dispersion relation in ν for fixed q^2 (for $q^2 < 0$) (a once subtracted dispersion relation)

$$T_1(q^2,\nu) - T_1(q^2,0) = \frac{\nu^2}{2\pi}\int_0^\infty \frac{d(\nu')^2}{(\nu')^2(\nu'^2-(\nu+i\varepsilon)^2)}\,W_1(q^2,\nu') \tag{40.6a}$$

(Assuming no seagulls beyond a constant term absorbed in $T_1(q^2,0)$. This can be seen by direct substitution of (40.6) into (40.6a), remembering that $W(q^2,-\nu') = -W(q^2,\nu')$, i.e. that $h(\sigma,\beta) = -h(\sigma,-\beta)$.

Consider

$$\int \frac{d\nu'}{\nu'-(\nu+i\varepsilon)}\,\delta(q^2+2M\beta\nu'-\sigma)\,\mathrm{sgn}(\nu'+M\beta) = \frac{-1}{q^2-\sigma+2M\beta\mu+i\varepsilon\,\mathrm{sgn}\beta},$$

everything is clear except the $i\varepsilon\,\mathrm{sgn}\beta$ should be $i\varepsilon\,\mathrm{sgn}(\nu+M\beta)$. Let us choose to work with $\nu > 0$ (T is symmetric in ν). Then for $q^2 < 0$ there is no pole in $(q^2+2M\beta\nu-\sigma+i\varepsilon)^{-1}$ if β is negative, the sign of $i\varepsilon$ is arbitrary; we may replace $\mathrm{sgn}\beta$ by $\mathrm{sgn}(\nu+M\beta)$. If $\beta > 0$, $\mathrm{sgn}\beta = + = \mathrm{sgn}(\nu+M\beta)$ so it is still correct. (For ν negative the sign reverses.) Hence we have

$$(T_1(q^2,\nu) - T_1(q^2,0))/\nu = -\int_{-\infty}^{\infty} \frac{d\nu'}{\nu'-(\nu+i\varepsilon)}\,\frac{W_1(q^2,\nu')}{\nu'}$$

W_1/ν' is symmetric so we can set the range from 0 to ∞ with $1/(\nu'-(\nu+i\varepsilon)) + 1/(-\nu'-(\nu+i\varepsilon)) = -2\nu/(-\nu'^2+(\nu+i\varepsilon)^2)$. From this equation (40.6a) follows.

For W_2 which converges faster we expect a corresponding unsubtracted dispersion relation

$$T_2(q^2,\nu) = \frac{1}{2\pi}\int_0^\infty \frac{d(\nu')^2}{(\nu')^2-(\nu+i\varepsilon)^2}\, W_2(q^2,\nu') \qquad \text{for } q^2 < 0 \ .$$

For $q^2 = 0$ these are of course the Kronig-Kramers formulas relating real and imaginary parts of the index of refraction (forward light scattering) which is necessary if signals are not to come out before they go into a block of scattering material.

We try now once again (unsuccessfully as it will turn out) to try to obtain some expected limitations on $2MW_1$ because we know it is a causal (zero outside the light cone in space) commutator. This time we shall use behavior valid not only for large ν but for any region. For example we know for the elastic scattering if the proton were a point charge we would have for $\nu > 0$, $2MW_1 = 2M\nu\delta((p+q)^2-M^2) = 2M\nu\delta(q^2+2M\nu)$. (To get the correct symmetry for ν positive and negative this can be written more correctly as $2MW_1 = 2\,M\nu(\mathrm{sgn}(\nu+M)\,\delta(q^2+2M\nu) - \mathrm{sgn}(\nu-M)\delta(q^2-2M\nu))$. This is obviously causal - as we have seen - for it comes from perturbation theory of fields.

Now in the real world this is multiplied by a factor (the square of the elastic form factor), a function of q^2, say $f(q^2)$ which falls off gradually, for negative q^2, from $q^2 = 0$, to behave as $(-q^2)^{-\gamma}$ for large $-q^2$. One would expect such modulation of the function expected for point-like particles to represent some kind of smearing of the point and to perhaps imply a lack of causality - a lack which must be balanced by correct contributions from terms off the elastic mass shell (corresponding to other resonances, etc). Thus, can we not make some requirements of behavior in other regions, especially non-elastic regions by our knowledge of $f(q^2)$ for negative q^2? Unfortunately not (as Cornwall, Corrigan and Norton, (Phys. Rev. D3 537 (1971)) show) it is possible to stay on the elastic mass shell for negative q^2 and only alter the behavior for positive q^2 and still arrange virtually any $f(q^2)$ which falls off fairly smoothly. In particular if $f(q^2)$ can be written in the form

$$f(q^2) = \int_0^\infty \frac{\rho(\mu)\, d\mu}{q^2-\mu} \tag{40.7}$$

for negative q^2, it can be done. We can see why this is, and also get a better physical feel for constructing causal functions by the following considerations. It is easiest to deal with the scattering functions T, for these to be causal they must be the Fourier transform of a retarded commutator and thus zero outside a forward light cone. Let $a(x,t)$ be such a function and $A(q)$ its F.T. and let $b(x,t)$ be another such function, $B(q)$ its F.T. Thus the convolution of $a(x,t)$ and $b(x,t)$ is obviously, by geometry, such a function (zero outside forward light cone) and hence its F.T. or simply $A(q)B(q)$ is again satisfactory (a causal scattering function).

We see that combinations by multiplication (and addition of causal scattering functions are causal scattering functions. The simplest causal scattering function is

$$[q^2 - m^2 + i\varepsilon \operatorname{sgn} q_4]^{-1} \tag{40.8}$$

We can generalize this (multiply by $e^{iu\cdot x}$ in space time) to find for any four-vector u_μ that

$$[(q+u)^2-m^2 + i\varepsilon \operatorname{sgn}(q_4+u_4)]^{-1} \tag{40.9}$$

is a causal function. Thus the elastic scattering function for point particles

$$[(q+p)^2-M^2 + i\varepsilon \operatorname{sgn}(\ +M)]^{-1} \tag{40.10}$$

evidently is causal (as well as what you get by putting $-q$ for q). We can multiply this by an expression like (40.8), we see

$$[q^2-m^2+i\varepsilon \operatorname{sgn} \nu]^{-1}\ [(q+p)^2-M^2 + i\varepsilon \operatorname{sgn}(\nu+M)]^{-1}$$

is causal. This is true for any $m^2 = \mu$ and hence any superposition with weight $\rho(\mu)d\mu$ is, so a scattering amplitude like (note $(q+p)^2-M^2 = q^2+2M\nu$)

$$\frac{T_1}{\nu} = \int \frac{\rho(\mu)d\mu}{q^2-\mu+ i\varepsilon \operatorname{sgn}\nu}\ \frac{1}{q^2+2M\nu+ i\varepsilon \operatorname{sgn}(\nu+M)} \tag{40.11}$$

is, by itself causal. To get the correct symmetry for ν one need only add the corresponding expression with ν replaced by $-\nu$; we will suppose it is always done and not write it explicitly.

(You might find it physically more satisfactory and easier to interpret if the factor is considered as a form factor due to virtual particles like ρ's at each vertex. Those at one vertex would contribute a factor

$$g(q^2) = \int_0^\infty \frac{w(\mu)d\mu}{q^2-\mu+ i\varepsilon\ \mathrm{sgn}\nu} \tag{40.12}$$

where μ is the mass squared of the virtual particle and w measures the weight of its contribution; g is causal. We would then expect to multiply (40.10) by $(g(q^2))^2$, one factor for each coupling; the result would still be causal, and possibly physically easier to understand.)

To get W_1/ν (commutator) from (40.11) we need only take its imaginary part

$$\frac{W_1}{\nu} = \int_0^\infty PV \frac{1}{q^2-\mu}\ \rho(\mu)d\mu\ \pi\ \mathrm{sgn}(\nu+M)\delta(q^2-2M\nu) +$$

$$+\int_0^\infty \pi \mathrm{sgn}\nu\delta(q^2-\mu)\rho(\mu)d\mu\ PV\ \frac{1}{q^2+2M\nu} \tag{40.13}$$

In the region $q^2 < 0$ the last term disappears and we are just left with the elastic point charge scattering multiplied by a factor $f(q^2)$ given by (40.7).

It is disappointing that the restrictions of causality do not affect our region of experimental observation ($q^2 < 0$). Furthermore, it will be hard in practice to get $\rho(\mu)$ exactly from knowledge of the integral ($-q^2 = Q^2$)

$$f(Q^2) = \int_0^\infty \frac{\rho(\mu)}{(Q^2+\mu^2)}\ d\mu \tag{40.14}$$

even if fairly exact. It is not easy to reverse the integral into accurate knowledge of $\rho(m)$ unless physical arguments (ρ dominance etc.) are available. Thus again we are thrown back in this problem to understand the process physically; the mathematical properties do not help as much as we had hoped.

(We know that if $f(Q^2)$ falls faster than $1/Q^2$, say as $(1/Q^2)^4$; then we can conclude, by (40.14), that $\int_0^1 \rho(\mu)d\mu = 0$. Such a relation is called a superconvergence relation $-f(Q^2)$ converges for large Q^2 more rapidly than the form would suggest. Again since f falls as $(1/Q^2)^4$ we can conclude that the moments $\int_0^\infty \mu^n \rho(\mu)d\mu$ vanish for n = 0, 1, 2. Alternatively, f can be expressed as $(g(q^2))^2$ with $g(q^2) = \int \gamma(\mu)(q^2-\mu)^{-1}d\mu$ and $\int \gamma(\mu)d\mu = 0$.)

Lecture 41

The idea discussed at the end of the previous lecture is immediately generalizable to scattering through an intermediate resonance, say of mass M_x^2, call $M_x^2 - M^2 = \lambda$. A point coupling would give scattering as $(q^2+2M\nu - \lambda + i\varepsilon\ \mathrm{sgn}(\nu+M))^{-1}$. We can multiply by any form factor, say with a $\rho(\lambda,\mu)$. We are thus representing things by a sum of s channel resonances each of which has a square forr factor

$$f(\lambda,-q^2) = \int \frac{\rho(\lambda,\mu)}{-q^2+\mu}\,d\mu \tag{41.1}$$

The total scattering from all this (s channel resonance representation) would then be

$$\frac{T_1}{\nu} = \int_0^\infty\int_0^\infty \frac{\rho(\lambda,\mu)}{q^2-\mu+i\varepsilon\ \mathrm{sgn}\nu}\ \frac{d\lambda d\mu}{q^2+2M\nu-\lambda+i\varepsilon\ \mathrm{sgn}(\nu+M)} +$$

$$+ \text{ same term } (\nu \to -\nu) + \text{seagulls} \tag{41.2}$$

The W_1 which goes with this (the imaginary part of T_1/ν) is

$$\frac{W_1}{\nu} = \int_0^\infty\int_0^\infty \rho(\lambda,\mu)\ \mathrm{PV}\frac{\pi}{q^2-\mu^2}\,[\mathrm{sgn}(\nu+M)\delta(q^2+2M\nu-\lambda) - \mathrm{sgn}(\nu-M)\delta(q^2-2M\nu-\lambda)] +$$

$$+ \pi\,\mathrm{sgn}\nu\int_0^\infty\int_0^\infty \rho(\lambda,\mu)\delta(q^2-\mu^2)\ \mathrm{PV}\left(\frac{1}{q^2+2M\nu-\lambda} - \frac{1}{q^2-2M\nu-\lambda}\right) \tag{41.3}$$

For $q^2 < 0$ the last term vanishes and we have for $\nu > 0$ simply

$$W_1/\nu = \pi\int_0^\infty f(\lambda,-q^2)\delta(q^2+2M\nu-\lambda)d\lambda = f(q^2+2M\nu,\ -q^2) \tag{41.4}$$

a superposition of contributions for each effective mass with a form factor $f(\lambda, -q^2)$ of form given in (41.1).

Have we not gone in a complete circle? Our original expression giving (except for photon polarization factors) W_1 was of the form

$$\sum |\langle p|J(Q)|x\rangle|^2\ \delta((q+p)^2-M_x^2) \qquad (\text{for } q^2 > 0,\ \nu > 0)$$

This looks just like (41.4). The δ is of course $\delta(q^2+2M\nu-\lambda)$ so we interpret $f(\lambda, -q^2)$ as $\sum |\langle p|J(Q)|x\rangle|^2$ summed over states of a given mass^2 = $M^2+\lambda$. (At first this would seem to make a function of Q^2, the space part of the momentum transfer squared, instead of $q^2 = \nu^2 - Q^2$; but it is the same because the δ function relates ν to Q^2 and it can be expressed either way.) We see $f(\lambda, -q^2)$ must be positive for $-q^2 > 0$; of course, since the lowest elastic $M_x = M$ ($\lambda = 0$) is separated from the continuum at $M_x = M+m_\pi$ ($\lambda_{th} = 2m_\pi M + m_\pi^2$). The function $f(\lambda, q^2)$ and hence $\rho(\lambda,\mu)$ will have a $\delta(\lambda)$ contribution and after that the integral in (41.3) will start at an inelastic threshold λ_{th} to ∞.

We may not have gone in a complete circle. First we now know (a) that the weight factor $f(\lambda, -q^2)$ must be expressible in the form (41.1) and (b) we know what the function looks like in the experimentally unavailable region $+q^2 > 0$ (see 41.3) and, of course, the scattering function (41.2) that goes with it. But do we know this?

We only proved that the form (41.2) was causal; we have not proven that every causal function must be expressible as (41.2) and at the moment we do not think we can.

Since (41.2) is causal it must be expressible in the DGS form (39.6). One way to do this (suggested by Cornwall, Corrigan and Norton, Phys. Rev. D3 536 (1971)) is to combine denominators (it is easier to use the Feynman amplitudes replacing $i\varepsilon\,\mathrm{sgn}(..)$ by $i\varepsilon$) via

$$\int_0^1 \frac{d\beta}{[q^2+2M\nu\beta-\lambda\beta-\mu(1-\beta)]^2} = \frac{1}{q^2-\mu}\ \frac{1}{q^2+2M\nu-\lambda}$$

to get a single denominator. Thus we call $\sigma = \lambda\beta+\mu(1-\beta)$ and integrate on σ by parts to prove

$$h(\sigma,\beta) = \int_0^\infty\int_0^\infty \rho(\lambda,\mu)\,\delta'(\lambda\beta+\mu(1-\beta)-\sigma)\,d\lambda\,d\mu \tag{41.5}$$

Of course this can now be simplified. If from $h(\sigma,\beta)$ we could always find a

$\rho(\lambda,\mu)$ which would give this h we would have a proof (assuming the DGS representation is proved) that (41.2) is also a necessary form. We suspect it cannot be done and it is possible that (41.2), although very physical, is not the complete expression; but other forms (other types of diagrams - or "channels" other than s channels) might have to be added to it to get a complete representation. This is a good problem.

W for all q^2 in terms of W for $q^2 < 0$?

We have a form that suggests the question as to whether knowledge of the fact that W is causal, and knowledge of its value for negative q^2 only, enables us to find it for all q^2, ν. We are not now concerned with the practical fact that knowledge of $f(\lambda, -q^2)$ to a given limited experimental accuracy does not permit discovering $\rho(\lambda,\mu)$ very well and does not mathematically alone define $f(\lambda, -q^2)$ for $q^2 > 0$. Rather we suppose W_1 perfectly known for $q^2 < 0$ and ask to what extent T is defined everywhere.

This has great interest for there are quantities, such as electromagnetic self energies, which can be defined in terms of T as integrals perhaps (e.g. $\int T(q^2)d^4q/q^2$). If T were uniquely determined by W in the experimentally accessible region we might search for expressions for these integrals directly in terms of this W at $q^2 < 0$. (Cottingham formula for self-energy.)

Given $W(q^2,\nu)$ in the available region (momentum like q^2) how unique is $W(q^2,\nu)$ in the energy-like region? Let $W_a(q^2,\nu)$ and $W_b(q^2,\nu)$ be two causal functions identical in the $q^2 < 0$ region. Let us study what is possible for their difference $W_d = W_a - W_b$. $W_d(q^2,\nu) = 0$ for $q^2 < 0$ and hence is causal. What form must it have? We see immediately it need not be zero because $\mathrm{sgn}\nu\,\delta(q^2-m^2)$ is such a function. To get the most general form we use Dyson's theorem which says (since it is causal) that it must be expressible in the form

$$W_d = \int d^4u \int_0^\infty ds^2\ \mathrm{sgn}(q_o-u_o)\,\delta((q-u)^2-s^2)\,\Phi(u,s^2) \tag{41.6}$$

where Φ is non-zero inside the region where the hyperboloid $(q-u)^2=s^2$ does not penetrate the region S of q space where we know W_d vanishes. Thus S is the entire $q^2 < 0$ region. It is seen that every hyperboloid cuts S unless u = 0. Thus the most general form for W_d is

$$W_d = \int_0^\infty \text{sgn}\nu\,\delta(q^2-s^2)\phi(s^2)ds^2 = \text{sgn}\nu\,\phi(q^2) \tag{41.7}$$

where by definition $\phi(x) = 0$ for $x < 0$.

Thus complete knowledge of $W_1(q^2,\nu)$ in the experimentally available $q^2 < 0$ region permits definition of W_1 everywhere within an arbitrary constant, (independent of ν) function of q^2 for positive q^2. T is determined also up to the function

$$\int \frac{\phi(s^2)ds^2}{q^2-s^2+i\varepsilon\ \text{sgn}\nu} \tag{41.8}$$

Strictly this argument is not valid, gradients of δ functions in space time could be used in Dyson's theorem. More correctly, at each positive q^2 the function W is determined by the behavior of $W(q^2,\nu)$ for $q^2 < 0$ up to an unknown polynomial in ν, the coefficients of which are arbitrary functions of q^2. Physical arguments about large ν asymptotic behavior would have to be used to limit the degree of these polynomials. The scaling limit for $x < 0$ and the fact that W is odd tells us that (41.7) must be of the form $\nu\phi(q^2)$ (for $\nu > 0$) and that T_1 is determined up to a function of q^2 given by (41.8). Note this function is not zero for $q^2 < 0$ so T_1 is not completely determined for $q^2 < 0$ by W. This agrees with the dispersion result (40.6a) where a subtraction had to be made and an arbitrary function $(T_1(q^2,0))$ left undetermined.

Electromagnetic Self Energy

Lecture 42

Electromagnetic Self Energy

We now discuss a few places where knowledge of T would help us calculate the electrodynamic properties of protons and neutrons. Since we have measured, in W, the electromagnetic coupling of protons we might hope to use the experimental knowledge to determine the electromagnetic energy of proton and neutron and their measured difference, to compare to experiment. As we shall see the hope is, at present, frustrated because knowledge of W for $q^2 < 0$ where it is available is not quite enough to determine the electromagnetic coupling everywhere (T) - the arbitrary function of $\phi(s)$ mentioned in the previous lecture is not determined. Since the answer for the p-n mass difference is only one number we are frustrated until we can find a theoretical or experimental way to determine T uniquely - for example to determine $T_1(q^2,0)$ of the dispersion relation (40.6a) for $q^2 < 0$.

Before we discuss this by formal mathematics let us see what we can expect. As is well known, the self mass of a point spin 1/2 particle diverges logarithmically. For the electron $\Delta(m^2) = m^2 \frac{3e^2}{2\pi} \ln \frac{\Lambda^2}{m^2}$ where Λ is some upper cut off for electrodynamics if we replace the photon propagator

$$\frac{1}{q^2} \to \frac{1}{q^2}\left(\frac{-\Lambda^2}{\Lambda^2+q^2}\right)$$

This Δm^2 is experimentally undefinable. A similar infinity for the ΔM^2 electromagnetic for the proton would also - by itself not be observable, but

$$(\Delta M^2)_{proton} - (\Delta M^2)_{neutron} = -1.2934 \text{ MeV}$$

is, in fact, observable and is measured to 5 significant figures. Can we calculate it - or even see what order of magnitude it is? - for example does our present theory say it must be infinite?

As long as the electromagnetic interaction of the nucleons were mysterious one could always say anything could happen - but now that we have some knowledge of them we must answer more specifically.

The divergence occurs from high frequencies and at first it was thought the hadrons might be soft at high frequency and the electromagnetic self energy convergent. But now we know for inelastic scattering at least they look like they are made of point-like constituents. Does this point-like behavior mean the energy must diverge? Of course, the proton could diverge and the neutron also - only the difference need converge - but the difference in point-like structure is $W_{1p}-W_{1n}$ is also finite and point-like in the scaling limit. It is the coincidence of the δ-function of the photon prepagator and of the electron propagator which makes the divergent self energy - and now we see the proton, the neutron, and proton minus neutron all have singular behavior on the light cone so it at first looks like divergence is inevitable.

Let us estimate how much. Since we are discussing the high energy behavior we can use the idea that proton=partons. Clearly the self energy diagram of most importance at high energy in the scaling limit are when a photon is emitted and absorbed by the same parton - therefore, as far as the divergent ($\ell n\ \Lambda^2$) part is concerned it is as if each parton gets a shift in mass proportional to $\Delta m_i^2 = m_i^2 e_i^2 \ell n \Lambda^2/m_i^2$ where e_i, m_i are the charge and mass of a parton. How much does this change the mass of the proton? We cannot honestly say. One might try to calculate the change in $E - P_z = M^2/2P$ by calculating the sum of the changes in $\varepsilon - p = \frac{Q^2+m^2}{2Px}$ of each parton. We would find

$$(\Delta M^2)_{nucleon} = \sum_{partons} \frac{\Delta m^2}{x} = \int \frac{m^2}{x} f^{ep}(x) \ell n\Lambda^2 dx$$

There are objections to this (by the way it is also even more divergent since $f_{ep} \sim 1/x$ and the x integral cannot be carried to 0). The energy is not just the sum of the kinetic energies of the partons, interaction energies among the partons are also involved. This is reflected in the dangerous formula containing the mass squared of a parton - a thing we said was meaningless to $\pm\Delta$ up to now. The distribution of partons is changed in first order also so the total m^2 change is not correctly evaluated. (It might be thought we are right to take $<\psi|\Delta V|\psi>$ for the perturbation on the Hamiltonian using the old wave function - but $\Delta V = \Delta m^2 a^* a$ is not the change in the Hamiltonian, only of the Lagrangian - in the Hamiltonian there are many other effects on the interaction terms through $1/\sqrt{2\omega}$ factors etc., so we have not computed the effect of the perturbation correctly.)

It may well be that $m_i^2 = 0$, or is effectively zero (a suggestion to me due to Zachariasen) due to interactions (or as a matter of principle) so the logarithmically divergent perturbations Δ_i depending on $\ln\Lambda^2$ never arise. This may be so, but we do not know. We must turn to more detailed quantitative analyses if we are to try to study this further.

Electromagnetic mass shifts come from the emission and reabsorption of a virtual photon.

QED tell us it is $\int <p|\{J_\mu(2)J_\mu(1)\}_T|p>\delta_+(s^2_{12})d\tau$ where $\delta_+(s^2_{12})$ is the propagator of the photon. Hence by taking Fourier transform

$$\Delta M^2 = 4\pi e^2 \int \frac{d^4q}{(2\pi)^4}\frac{1}{q^2+i\varepsilon}\; T^F_{\mu\mu}(q^2,\nu) \tag{42.1}$$

We have written

$$T_{\mu\nu} = (P_\mu - \frac{P\cdot q}{q^2}q_\mu)(P_\nu - \frac{P\cdot q}{q^2}q_\nu)T_2 - (\delta_{\mu\nu} - \frac{q_\mu q_\nu}{q^2})M^2 T_1$$

so that

$$T_{\mu\mu} = 4M^2(T_1 + [(1-\nu^2/q^2)T_2 - T_1]/4) \tag{42.2}$$

The expression in square brackets has the imaginary part $(1-\nu^2/q^2)W_2 - W_1$ which is the contribution from longitudinal photons. If partons are spin 1/2 it falls faster with ν than does T_1 whose imaginary part W_1 simply scales to $f(x)$ in the Bjorken limit. We will hereafter just write T for $T_1 + [(1-\nu^2/q^2)T_2 - T_1]/4$

and suppose the imaginary part $W = W_1+[1-\nu^2/q^2)W_2-W_1]/4$ scales to $f(x)$ in the Bjorken limit.

Lecture 43

Cottingham formula

As we have seen we can write for the electromagnetic mass effect

$$\frac{\Delta M^2}{4M^2 4\pi e^2} = I = \int \frac{d^4q(2\pi)^4}{q^2} \; T^F(q^2,\nu) \tag{43.1}$$

where $4T = T_{\mu\mu}$. We must integrate this over all q^2 but we have seen T is determined by its behavior for $q^2 < 0$ (where experiment can say something about it) so it is possible that maybe (43.1) can be written in terms of T for $q^2 < 0$ only. How this can in fact be done was shown by Cottingham. He showed that in the four-dimensional integral the contour on ν could be changed (without passing any singularities) from the real line $\nu = -\infty$ to ∞ to the imaginary axis $\nu = i\omega$, $\omega = -\infty$ to ∞. (We show how it is done later.) Suppose it is true. We can then write $d^4q = d\omega 2\pi Q dQ^2$, $q^2 = \nu^2-Q^2 = -\omega^2-Q^2$

$$I = \frac{1}{8\pi^3}\int \frac{Qd\omega dQ^2}{\omega^2+Q^2} \; T(-(\omega^2+Q^2), i\omega)$$

Now replace $-(\omega^2+Q^2)$ by $(-q^2)$, $Q = \sqrt{(-q^2)-\omega^2}$ to get

$$I = -\frac{1}{8\pi^3}\int_0^\infty \frac{d(-q^2)}{(-q^2)} \int_{-\sqrt{-q^2}}^{+\sqrt{-q^2}} \sqrt{(-q^2)-\omega^2}\; T(q^2,i\omega)d\omega \tag{43.2}$$

(Cottingham formula)

Now the quantity is all right for $-q^2$ but is completely unphysical for ν is imaginary. We can define it however by analytic continuation by our dispersion relation (40.6a), setting $\nu = i\omega$

$$T(q^2,i\omega) = T(q^2,0) - \frac{\omega^2}{\pi}\int_0^\infty \frac{d(\nu')^2}{\nu'^2(\nu'^2+\omega^2)} W(q^2,\nu') \tag{43.3}$$

Hence

$$8\pi^3 I = \int_0^\infty \frac{d(-q^2)}{-q^2}\int_{-\sqrt{-q^2}}^{+\sqrt{-q^2}} d\omega \sqrt{-q^2-\omega^2}\left\{T(q^2,0) - \frac{\omega^2}{\pi}\int_0^\infty \frac{d\nu'^2 W(q^2,\nu')}{\nu'^2(\nu'^2+\omega^2)}\right\} \tag{43.4}$$

We can carry out the integrals on ω to get

$$8\pi^3 I = \int_0^\infty \frac{d(-q^2)}{-q^2}\left\{\frac{\pi}{4}(-q^2)T(-q^2,0) - \frac{1}{2}\int_0^\infty\left(\sqrt{\frac{-q^2}{\nu'^2}+1} - 1 - \frac{1}{2}\frac{-q^2}{\nu'^2}\right)W(q^2,\nu')d\nu'^2\right\} \tag{43.5}$$

This then succeeds in getting an explicit formula for the self energy in terms of $W(q^2,\nu)$ in a region accessible to experiment. However, we are frustrated by the unknown term $T(q^2,0)$.

It is essential to know something about $T(q^2,0)$ if we are to be able even to determine whether the self energy diverges. Let us look first at the contribution from the second term in (43.5) in the scaling region from which divergences could come. Put $-q^2 = 2M\nu x$ and consider W as a function of x and ν, $W(x,\nu)$ which for large ν converges to the limit $f(x)$. The term becomes

$$\int_0^\infty \frac{dx}{x}\,\nu d\nu\left[\sqrt{\frac{2Mx}{\nu}+1} - 1 - \frac{1}{2}\left(\frac{2Mx}{\nu}\right)\right]W(x,\nu)$$

or for large ν, where the square bracket is $-(2Mx/\nu)^2/4$ we get

$$\int^\infty \frac{d\nu}{\nu} M^2 \int_0^1 xf(x)\,dx \quad .$$

The ν integral diverges logarithmically with coefficient $\int_0^1 xf(x)dx$, the fraction of momentum carried by the charged partons each weighed by the charge squared.

Of course a cut-off of electromagnetism is used in (43.5), $d(-q^2)/(-q^2)$ is replaced by $\frac{d(-q^2)}{-q^2}\;\frac{\Lambda^2}{\Lambda^2-q^2}$.

This provides a cut-off for our ν integral (of order $\Lambda^2/2Mx$) so the part diverging as $\ln\Lambda^2$ has a coefficient $\int xf(x)dx$.

However, the other term in $T(q^2,0)$ could also produce a $\ln\Lambda^2$ term if it only falls as $C/(-q^2)$ as $-q^2\to\infty$. (If $T(q^2,0) = \int\phi(s)ds/(q^2-s^2)$ this C is $\int\phi(s)ds$.) It is therefore possible that these divergences from T and W cancel and that the self energy (or at least the proton-neutron difference) is finite and calculable.

There are two views we could take at present. We know of course the p-n mass difference converges so let us talk about T and W for the p-n difference at least. In principle T could be determined by experiment and is therefore defined physically, either of the following could happen.

(a) Equation (43.5) with this T still gives a logarithmic divergence. The reason is that our theories are wrong for high energy; this and the electromagnetic self energy calculation of QED are both wrong and will both be fixed by the same modification of relativistic quantum mechanics at high energy some day to be found.

(b) The T is such that the integrals converge and agree with the experimental mass differences.

(c) The $T(q^2,0)$ (for $q^2 < 0$) is in fact not precisely definable experimentally, that it is to some extent arbitrary - hence that our theory is not able to calculate this mass difference precisely and must be "renormalized." I believe in this case, if partons are quarks only one renormalization constant, corresponding to the electromagnetic mass difference of u and d quarks, would suffice to make all the hadron self energy differences converge simultaneously.

As Zachariasen has suggested to me, the best (most limiting) thing to do today is to assume (b) is true. This puts restrictions on possible theories and may have predictive value. If it leads to a paradox or inconsistency we learn that (a) must be so. Zachariasen has shown that all will be convergent if the equal time commutator of J and $\dot{J}$ vanishes $[J_\mu, \dot{J}_\mu] = 0$. In the quark model it corresponds to quarks having zero rest mass.

I believe this is a very good problem to work on. I myself haven't found enough time while preparing these notes to analyze it in a more elementary or clear fashion for you.

How can we ever hope to get at $T(q^2,0)$ experimentally or theoretically? It is the forward scattering amplitude on a proton of a virtual photon of mass $-q^2$. It would be involved in a two-electron forward scattering e+e+p → e+e+p via the diagram

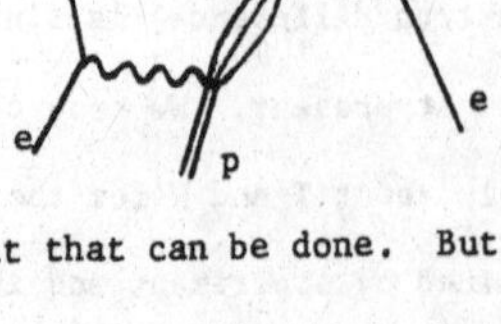

This is not an experiment that can be done. But knowledge of $T(q^2,\nu)$ anywhere

would be of assistance because the dispersion relations can be used to convert it to knowledge of $T(q^2,0)$; still no experiment suggests itself.

It is interesting that $T(0,0)$ can be obtained theoretically - the forward Compton scattering from a proton off a real (on shell) photon. For $Q \to 0$, $\nu \to 0$ we have very long wave length slow fields to which, of course, the proton looks to be simply a massive point charge of mass M. It scatters then as it would classically (or non-relativistically via Schrödinger equation from the $\frac{e^2}{2M}\vec{A}\cdot\vec{A}$ term) thus (called Rayleigh scattering) it gives

$$T(0,0) = -\frac{e^2}{M}.$$

NOTE: How to rotate contour to get Cottingham formula;

Use DGS representation for T^F

$$I = \frac{d^4q}{\nu^2-Q^2+i\epsilon}\;\frac{H(\sigma,\beta)}{\nu^2-Q^2+2M\nu\beta-\sigma+i\epsilon}$$

Call $E = \sqrt{\sigma+M^2\beta+Q^2}$ and note the singularities are at $\nu = Q-i\epsilon$, $-Q+i\epsilon$; $\nu+\beta M = E - i\epsilon$, $-E+i\epsilon$

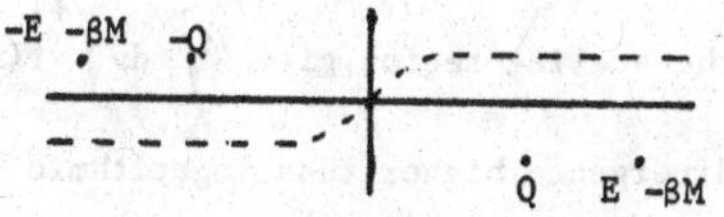

Since $E > \beta M$, the poles below the axis are all for $\nu > 0$. The contour goes like the dotted line, it can evidently be rotated to the imaginary axis.

Lecture 44

Expression for self energy in terms of W only

We didn't succeed in representing ΔM^2 in terms of $W(q^2,0)$ for negative q^2 only, without at the same time involving ourselves with another unknown function $T(q^2,0)$; and further each of two parts is infinite and it is hard to guess at the difference. Perhaps we should abandon this and take a last look at just the ΔM^2 expressed in terms of $W(q^2,\nu)$ for positive and negative q^2. It is given by

$$\Delta M^2 = \frac{e^2M}{\pi^2}\int_0^\infty\int_0^\infty \frac{Q}{(\nu+Q)}\, W(\nu,Q)\,dQ\,d\nu \qquad (44.1)$$

(This is obtained by expressing T^F in the form

$$T^F_{\mu\nu} = \int \left(\frac{K_{\mu\nu}(Q,\nu')}{\nu-\nu'+i\varepsilon} - \frac{K_{\mu\nu}(Q,\nu')}{\nu+\nu'-i\varepsilon} \right) \frac{d\nu'}{2\pi}$$

and inserting into (42.1). $W(\nu,\vec{Q})$ depends only on Q, the magnitude of $\vec{Q}$; and we have $d^3Q = 4\pi Q^2 dQ$).

To study the possible divergence of the integral at least, we go to large ν and in fact to the scaling region $-q^2/2M\nu = x$, hence $Q = \nu+Mx$. We can write (consider 2MW a function of x and ν, $2MW(x,\nu)$)

$$\Delta M^2 = \frac{e^2}{2\pi^2} \int_0^\infty d\nu \int_{-\nu/M}^{1} dx\ 2MW(\nu,x)\ \frac{\nu+Mx}{2\nu+Mx} \tag{44.2}$$

For large ν the upper limit for x is kinematically 1. 2MW approaches the function f(x) for positive x in the scaling limit. For negative x (positive q^2) it approaches as we have seen $-f(-x)$; note, however, that $-x$ is not kinematically limited to 1.

Dynamically $-x$ is limited, $W(x,\nu)$ exists for $x < -1$ but for large ν falls rapidly until there is nothing left of order one.

The contribution of the scaling region gives $\int_0^\infty d\nu \int_{-1}^{+1} f(x)dx = 0$.

But this only says that a divergence higher than logarithmic vanishes, a thing we expect anyway. Expanding the factor $(\nu+Mx)/(2\nu+Mx)$ seems to give a term $Mx/2\nu$ leading to $\int xf(x)(d\nu/\nu)dx$; this is not all, we shall have to know to order $1/\nu$ how W differs from its Bjorken limit for both positive and negative q^2.

This sums up the problem; experiment can in principle help us with the approach to the limit for q^2 negative. But we shall have to rely on theory to obtain the contributions to order $1/\nu$ for positive q^2 before we can decide whether ΔM^2 diverges according to present theories.

Other electromagnetic energies, Quark model

Having failed with fundamental theory to get information on electromagnetic mass differences, we now turn to much cruder pictures to discuss the possible relations of ΔM^2 in different terms of an SU_3 or SU_6 multiplet. We do it in the language of the quark model although many of the results come from weaker assumptions, like simple SU_3 etc.

The proton, for example, in the low energy quark model is made of three quarks, two u quarks and a d quark with total spin 1/2 with wave function

$$|p\rangle = \frac{1}{\sqrt{6}} \text{ (2 uud - udu - duu) } (\uparrow\uparrow\downarrow) \text{ symmetrized .} \qquad (44.3)$$

The electromagnetic self energy can be thought of as being made of two parts:

a) The self energy of the individual quarks. We suppose this is proportional to the change squared of each, thus 4a : a: a for u: d: s respectively. To a proton this contributes $\Delta M^2/2M_p = 9a$ (we shall normalize all mass squared changes to $2M_p$ of the proton as a scale for measuring a, the true change in M^2 is then $2M_p a$.)

b) An interaction energy between pairs which we take as proportional to the product of charges. The interaction must depend on the mutual spin relation of the pair. Thus write $\beta(1+\gamma)$ if spin is parallel and $\beta(1-\gamma)$ if spin is antiparallel; this is $\beta(1+\gamma P)$ where P is the spin exchange operator. We multiply by -2 for ud or us, +4 for uu and +1 for dd, ds, sd, ss pairs, call this factor x_{ij}.

The electromagnetic self mass operator can therefore be written

$$\Delta M^2 = 2\, M_p \,\{ 4a(\text{No. of u's}) + a(\text{No. of d's}) + a(\text{No. of s's}) +$$

$$+ \sum_{\text{pairs}} x_{ij}\ \beta(1+\gamma P)\} \qquad (44.4)$$

It is easy to get the expectation of this operator for every state. Thus on the proton the operator βx alone does not see the spins and gives $(2(4-2-2)\text{uud} - (4-2-2)\text{udu} - (4-2-2)\text{duu})\uparrow\uparrow\downarrow/\sqrt{6}$ which is zero. For the neutron, change a u to d, the coefficient is -3 so there is a contribution -3β. Next we study $\beta\gamma xP$

$$xP(\text{uud}\uparrow\uparrow\downarrow) = (4\text{uud} - 2\text{udu} - 2\text{duu})\uparrow\uparrow\downarrow$$

$$xP(\text{udu}\uparrow\uparrow\downarrow) = (-2\text{duu} - 2\text{uud} + 4\text{udu})\uparrow\uparrow\downarrow$$

$$xP(\text{duu}\uparrow\uparrow\downarrow) = (-2\text{udu} + 4\text{duu} - 2\text{udu})\uparrow\uparrow\downarrow$$

therefore

$$xP(2\text{uud} - \text{udu} - \text{duu})\uparrow\uparrow\downarrow/\sqrt{6} =$$

$$= ((8+2)\text{uud} + (-4-4+2)\text{udu} + (-4+2-4)\text{duu})\uparrow\uparrow\downarrow/\sqrt{6}$$

and

$$\langle p|xP|p\rangle = (20 + 6 + 6) = 6$$

Adding up the various contributions we have for the proton

$$\Delta M^2/2M_p = 9a + 6\beta\gamma$$

Lecture 45

Electromagnetic self mass, quark model (continued)

We can calculate the EM self masses for every particle of the $1/2^+$ octet as was done in the previous lecture for the proton, the result is:

$$p = 9a + 6\beta\gamma$$
$$n = 6a - 3\beta + 3\beta\gamma$$
$$\Sigma^+ = 9a + 6\beta\gamma$$
$$\Sigma^0 = 6a - 3\beta - 3\beta\gamma$$
$$\Sigma^- = 3a + 3\beta$$
$$\Xi^- = 3a + 3\beta$$
$$\Xi^0 = 6a - 3\beta + 3\beta\gamma$$
$$\Lambda^0 = 6a - 3\beta + \frac{3}{2}\beta\gamma$$

Hence

	ΔM_{exp}	$\Delta M^2/2M_p$
$p-n = 3a + 3\beta + 3\beta\gamma$	-1.29 MeV	-1.21 MeV
$\Xi^0 - \Xi^- = 3a - 6\beta + 3\beta\gamma$	$-6.6 \pm .7$	$-6.7 \pm .9$
$\Sigma^+ - \Sigma^0 = 3a + 3\beta + \frac{15}{2}\beta\gamma$	-3.06	-3.64
$\Sigma^0 - \Sigma^- = 3a - 6\beta - \frac{3}{2}\beta\gamma$	$-4.86 \pm .07$	-5.78
$(\Sigma^+ - \Sigma^-) - (p-n) - (\Xi^0 - \Xi^-) = 0$	$0 \pm .7$	$+.5 \pm .9$

We have three constants for four mass differences so we have the relation (an SU_3 relation)

$$(\Sigma^+ - \Sigma^-) - (p-n) - (\Xi^0 - \Xi^-) = 0$$

which fits well. (We choose to use ΔM^2 for no extremely good reason, the relation fits better with ΔM but there is little to choose, it is inside the experimental error for ΔM^2 also.)

The values of the constants that we get are

$3a = -2.0$ MeV

$\beta(1+\gamma) = .24$ MeV parallel spins

$\beta(1-\gamma) = 1.32$ MeV antiparallel spins

$\beta = .78$ MeV $\qquad \gamma = -.69$ MeV

The sign of a is the opposite of what you would expect but then you expect $+\infty$ and it must be renormalized, possibly to a negative value. It means generally particles with fewer u quarks are heavier, i.e. more positively charged baryons are lighter. The sign of the β term from electrostatic repulsion is positive as expected. We find attraction of parallel magnets in an s state but repulsion of antiparallel (the magnets are on top of each other) which is correct; the net repulsion in the parallel case is less.

Continuing in the 56 SU_6 multiplet to the decimet and supposing the constants are the same we can predict everything:

		$\Delta M^2/2M_p$ (predicted)
Δ^{++}	$= 12a + 12\beta(1-\gamma)$	-5.12 MeV
$\Delta^+ = \Sigma^+$	$= 9a$	-6.00
$\Delta^0 = \Sigma^0 = \Xi^0$	$= 6a - 3\beta(1+\gamma)$	-4.72
$\Delta^- = \Sigma^- = \Xi^- = \Omega^-$	$= 3a + 3\beta(1+\gamma)$	-1.28

	$\Delta M^2/2M_p$(pred.)	ΔM(pred.)	ΔM(exp.)
$\Delta^0-\Delta^{++}$	$+0.4$	$+0.3$	$2.9 \pm .9$
$\Delta^--\Delta^{++}$	$+3.8$	$+2.9$	7.9 ± 6.8
$\Delta^+-\Delta^{++}$	-0.9	-0.7	
$\Sigma^--\Sigma^+$	$+4.7$	$+3.2$	3.3 ± 1.5
$\Xi^--\Xi^0$	$+3.4$	$+2.2$	4.9 ± 2.0

The experimental data is not good but there is a serious discrepancy with a very recent experiment on $\Delta^0 - \Delta^{++}$; except for this the signs and general order of sizes is good.

For pseudoscalar mesons, noting that the sign of the charges is reversed for antiparticles, and that only antisymmetric spin contributes we get (call $\beta' = \beta(1-\gamma)$)

		$\Delta m^2/2M_p$ (exp.)
$\pi^+ = 5a + 2\beta'$		
$\pi^0 = 5a - \frac{5}{2}\beta'$	$\pi^+-\pi^0 = \frac{9}{2}\beta'$	.64
$K^+ = 5a + 2\beta'$	$K^+-K^0 = 3a + 3\beta'$	-1.95
$K^0 = 2a - \beta'$		

We have two constants to fit with two parameters and have no prediction. We get $\beta' = .14$, $3a = -2.38$ again confirming that a is negative. In fact a is close to its value for the baryons which is what FKR's relativistic quark model would expect. Furthermore the $\beta' = \beta(1-\gamma)$ is nearly of the same order of magnitude.

Finally we do the vector mesons. Here we need $\beta(1-\gamma) = b$ for parallel spins

$$\rho^+ = 5a + 2b$$
$$\rho^o = 5a - \frac{5}{2}b$$
$$\omega^o = 5a - \frac{5}{2}b$$
$$K^{*+} = 5a + 2b$$
$$K^{*o} = 2a - b$$

The only data on ΔM is that for $K^{*+} - K^{*o}$ which is -5.7 ± 1.7 MeV or $\Delta m^2/2M_p = -5.1 \pm 1.5 = 3a + 3b$. If $3a = -2.38$ this gives $\beta = -.9 \pm 5$, which is very bad, it is of the wrong sign. If the values for the octet baryons are used we predict $\Delta m^2/2M_p = \pm 1.9$!

In this system we also have two other effects (discussed in Lecture 15) a) the electromagnetic mixing matrix between ρ^o and ω^o, with off diagonal term $3a - 3b/2$. b) The annihilation term between ρ^o and ω^o which we found to be

$$\Delta m = \begin{array}{c} \\ \rho^o \\ \omega^o \end{array} \begin{array}{c} \rho^o \quad \omega^o \\ \begin{pmatrix} 1.53 & .51 \\ .51 & .17 \end{pmatrix} \end{array} \text{MeV.}$$

The off diagonal element δ in the mass matrix

$$\Delta m = \begin{pmatrix} m_\rho - i\Gamma_\rho/2 & \delta \\ \delta & m_\omega - i\Gamma\omega/2 \end{pmatrix}$$

is determined by ρ,ω interference to be $-3.7 \pm .9$ MeV. Subtracting the annihilation term +.51 gives $-4.2 \pm .9$ MeV., for the contribution of the self energy. $\Delta m^2/2M_p$ then corresponds to $3a - 3b/2 = 3.7 \pm .7$. This suggests if $3a = -2$ MeV that b is in fact positive and near $+ 1.1 \pm .4$ MeV, not too consistent with the baryon value of .24 for parallel spins.

<u>To summarize:</u> This is generally unsuccessful. For the baryons SU_3 works but SU_6 does not, predicting $\Delta^o - \Delta^{++} = + 0.3$ for the experimental $2.9 \pm .9$. For the mesons things are very poor. For pseudoscalars we get $3a = -2.4$, $\beta' = .14$ (compared to 1.3 for baryons). For the vector mesons the situation is confused.

If we use mass differences instead of $\Delta m^2/2M_p$ for the rules, the baryon constants come out $3a = -1.9$, $\beta(1+\gamma) = 0.20$, $\beta(1-\gamma) = 0.98$. The predicted

$\Delta^o - \Delta^{++}$ is 0.8 so only little is gained here. But the meson situation is altered (because of the small π mass). The constants come out for the pseudoscalars $3a = 7.0$, $\beta' = 1.0$. For the vector mesons we have $3a = -4.8$, $b = -0.3$.

The $\beta(1+\gamma)$ for baryons may differ from b for vector mesons and $\beta(1-\gamma)$ for baryons may also differ from β' for pseudoscalar mesons because the size of the wave function is so different so the mean 1/r differs. The electric and magnetic interaction need not change in the same ratio so it is possible that b is negative for mesons and positive for baryons, but it is difficult to see why β' is so different as the Δm^2 gives.

Why the value of 3a should differ in one case from the other is not clear. Were it not for experiment I would decide 3a calculated for $\Delta m^2/2M_p$ is the same for pseudoscalar and pseudovector mesons and is 0.6 of that for baryons. This is because we might guess that mass squared shifts due to strangeness come from a mass change in the s quark. The difference in m^2 for the baryons is about .40, for the mesons about .24 or 0.6 as much. Thus the self energy correction effect - if it is to be associated simply with a proper mass change of the u quark will also be 0.6 as effective in mesons as in baryons.

Evidently the naive theory does not work well, we do not understand things so well. A more detailed dynamical theory is necessary. But in any case, most particularly the large $\Delta^o - \Delta^{++}$ is most disquieting.

<u>$\Delta I = 2$ mass differences</u>

There is one observation that can be made here which suggests that dynamic calculations might be possible for some combinations. Notice that a is the self energy term, possibly involving high frequency behavior, but $\beta(1\pm\gamma)$ is due to mutual interaction and ought to show no divergence. Certain combinations of mass squared differences do not involve a. They are

$$(\pi^+ - \pi^o),$$

$$\frac{1}{2}(\Delta^{++} + \Delta^- - \Delta^o - \Delta^+) = (\Delta^+ + \Delta^- - 2\Delta^o),$$

$$\Sigma^+ + \Sigma^- - 2\Sigma^o$$

All these involve the $\Delta I = 2$ isospin part of the electromagnetic self energy. This energy depends on the products of two current operators JJ each of which contains $\Delta I = 0$ or 1, and can make terms whose parts are $\Delta I = 0$, 1 or 2.

In terms of I_z these go as constant, I_z, and $I_z^2 - \frac{1}{3} I(I+1)$ respectively. The constant ($\Delta I = 0$) part is experimentally lost in the strong interactions, the $\Delta I = 1$ or I_z terms are measured by differences proportional to I_z, like $\Sigma^+ - \Sigma^-$ or p-n. But the $\Delta I = 2$ term effect is proportional to I_z^2 and is measured by the differences (like $\Sigma^+ + \Sigma^- - 2\Sigma^o$) mentioned above.

We are calculating the $\Delta I_z = 0$ component of the $\Delta I = 2$ effect, of course, but to isolate it think of calculating an artificial $\Delta I_z = +2$ that would arise if electromagnetic current J were $\Delta I_z = +1$ (instead of $\Delta I_z = 0$), $\Delta I = 1$. We must then think of calculating an effect of two currents like $J^+ J^+$.

Now we see why a does not arise, and why in general, possibly the integrals in calculating this might converge rapidly at high ν, provided partons are quarks. For if all the current carrying fundamental operators (partons) have isospin 1/2 it is impossible for two J^+ to operate on the same parton in succession at high energy. Thus the virtual photon exchange diagrams like A contribute only to $\Delta I = 1$ and are impossible for $\Delta I = 2$, only B are allowed for $\Delta I = 2$

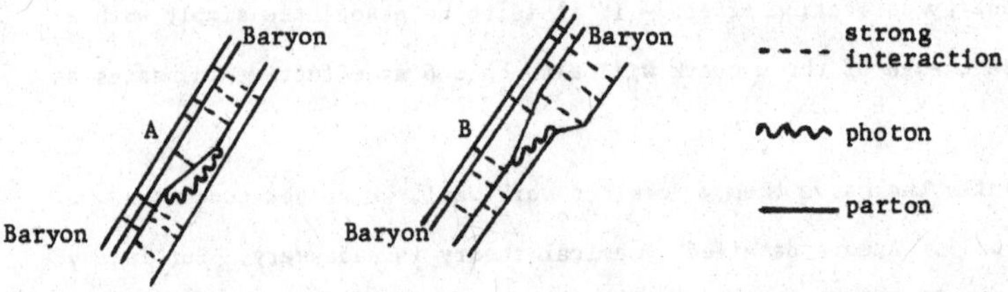

Large virtual momenta of the photon is possible in A no matter how soft the strong interactions are - if the initial momentum distribution of the partons involves only slow ones so can the final state. But in B, if the virtual photon momentum is very high it is unlikely that the soft strong interaction can pull the partons back together to give much of a diagonal amplitude to be in the undisturbed baryon state again.

Lecture 46

Further comments on electromagnetic mass differences

As we found in the previous lecture, dynamic calculations of $\Delta I = 2$ mass differences should be feasible by summing over not many states and using all we know theoretically and experimentally about the expected behavior of the necessary matrix elements. Of particular interest would be a study of the $\pi^+ - \pi^0$ mass difference.

The elastic term contributes almost all of the $\pi^+ - \pi^0$ mass difference as a simple calculation will show. The diagrams are

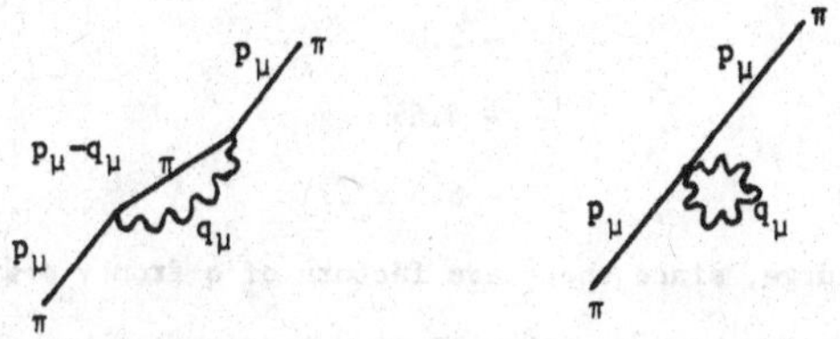

The pion form factor is dominated by the ρ resonance, we therefore include a factor $m_\rho^2/(q^2-m_\rho^2)$ at each photon coupling so the mass difference is

$$\Delta m^2 = 4\pi e^2 \int \left[\frac{(2p_\mu - q_\mu)(2p_\mu - q_\mu)}{(p-q)^2 - m_\pi^2} - \delta_{\mu\mu} \right] \frac{d^4q/(2\pi)^4}{q^2} \frac{m_\rho^4}{(q^2 - m_\rho^2)^2} \qquad (46.1)$$

To order zero in m_π^2/m_ρ^2 the first term in the square brackets is 1, this gives

$$(\Delta m^2)_0 = -4\pi e^2 \int 3 \frac{d^4q}{(2\pi)^4} \frac{1}{q^2} \frac{m_\rho^4}{(q^2 - m_\rho^2)^2}$$

The integral is readily performed remembering that in four dimensions $d^4q = \pi^2 q^2 dq^2$ after integrating over angles. The result of the zeroth order calculation is $(\Delta m^2)_0 = 3e^2 m_\rho^2/4\pi$, doing the next order gives

$$\Delta m^2 = \frac{3e^2 m_\rho^2}{4\pi} \left[1 + \frac{m_\pi^2}{m_\rho^2} \ln \frac{m_\rho^2}{m_\pi^2} + O\left(\frac{m_\pi^2}{m_\rho^2}\right)^2 \right] \qquad (46.2)$$

$$\approx 1.16 \times 10^{-3}\ (\mathrm{GeV})^2$$

In terms of $\Delta m = \Delta m^2/2m$ this is $\pi^+ - \pi^0 = 4.1$ MeV as compared to the experimental 4.6 MeV. Estimates of contributions of higher intermediate states (which could be done using FKR's model) should be small.

A similar result is found for $\Sigma^+ + \Sigma^- - 2\Sigma^0$. It seems as though the elastic term already gives almost the complete result. Estimates of higher resonances give less than 20%, and the result is in good agreement with the data.

The contributions of the elastic term to self energy differences of baryons were calculated by Gross and Pagels, Phys. Rev. 172 1381 (1968) using SU_3 for the magnetic moments, and G_E and G_M varying like $(1 + q^2/.71)^{-2}$. They get

	ΔM(elastic)	ΔM(exp.)
p-n	+ 0.79 MeV	- 1.29 MeV
$\Sigma^+ - \Sigma^0$	+ 0.16	- 3.06
$\Sigma^0 - \Sigma^-$	- 0.88	- 4.86
$\Xi^0 - \Xi^-$	- 1.10	- 6.5 ± .7

(Most of this is due to the charge, since there are factors of q from $\gamma_\mu \not{q} - \not{q}\gamma_\mu$ in the magnetic part.) Note that $\Sigma^+ + \Sigma^- - 2\Sigma^0 = +1.54$ MeV (elastic) compared to +1.8 experimental, as we expected this should be dominated by the elastic term. (Decimet intermediate states give less than .1 and presumably higher ones even less.)

Note that other combinations (other than $\Delta I = 2$) do not involve "a" (see lectures 44 and 45, "a" is the EM self energy of a quark) such as (p-n) - $(\Xi^0 - \Xi^-)$. In this case one cannot show that these differences also involve a product of currents such that each current does not act on the same quark. But the fact that "a" is not involved suggests that we may be able to get at this difference also by estimating matrix elements to various known states using, for example, the quark model. However, the elastic term fails utterly this time $\Delta m(\text{elastic}) = +1.9$, $\Delta M\text{exp.} = 4.2 \pm .7$. Why? To explain this we look at the contributions from the scaling region. Let u(x), $\bar{u}(x)$, d(x), $\bar{d}(x)$, s(x), $\bar{s}(x)$ be as described in lecture 31, then $\nu W_2/x = f(x)$ for p, n, Ξ^0 and Ξ^- is

$$f^{ep} = \frac{4}{9}(u + \bar{u}) + \frac{1}{9}(d + \bar{d}) + \frac{1}{9}(s + \bar{s})$$

$$f^{en} = \frac{1}{9}(u + \bar{u}) + \frac{4}{9}(d + \bar{d}) + \frac{1}{9}(s + \bar{s})$$

$$f^{e\Xi^0} = \frac{1}{9}(u + \bar{u}) + \frac{4}{9}(d + \bar{d}) + \frac{1}{9}(s + \bar{s})$$

$$f^{e\Xi^-} = \frac{1}{9}(u + \bar{u}) + \frac{1}{9}(d + \bar{d}) + \frac{4}{9}(s + \bar{s}) \tag{46.3}$$

(The neutron is obtained from the proton by replacing u by d and vice versa. The Ξ^0 is like the neutron but s replaces u and vice versa. The Ξ^- is like the Ξ^0 but u replaces d and vice versa.) Therefore the scaling function for (p-n)-

$(\Xi^0 - \Xi^-)$ is $\frac{1}{3}(u + \bar{u} + s + \bar{s} - 2d - 2\bar{d})$ which is not necessarily zero so high frequencies can come in. In the model of valence quarks plus sea it is zero but we do not believe this to be likely.

For further details on all these matters of EM self energy see an article by W.N. Cottingham in "Hadronic Interactions of Electrons and Protons", Cummings and Osborn Ed. Academic Press, N.Y. 1971.

Lecture 47

Compton effect $\gamma p \to \gamma p$ or $\gamma n \to \gamma n$

We now go on to discuss other effects involving two photon couplings. The Compton effect is the most closely related to what we have done. If the scattering is exactly in the forward direction the scattering amplitude is given by $T_{\mu\nu}(q^2,\nu)$ for $q^2 = 0$. We previously meant the average over proton spins, thus T is the spin averaged forward scattering, we could also measure for special spin directions of the proton. The imaginary part of the forward scattering is, of course, the total cross section $\sigma_{\gamma p}$ or $\sigma_{\gamma n}$ which we have discussed before. (E.g. $\sigma_{\gamma p}$ showed resonances at low ν, a fall off perhaps like $(97 + 67/\sqrt{\nu})\mu b$, and $(97 + 43/\sqrt{\nu}/\mu b)$ for neutrons.

The differential cross section can be fitted with

$$\frac{d\sigma}{dt} = \left(\frac{d\sigma}{dt}\right)_0 e^{At}$$

For energies from 2 to 7 GeV we get around 6 GeV^{-2}. For energies from 8 to 16 GeV, A = 8 GeV^{-2} is closer; there is some sign of a quadratic term $At + Bt^2$ with A = 7.4, B = 2.0. (This is much like hadron diffraction scattering, e.g. πp at 9 GeV has A = 9, B = 2.5.) Thus photon diffraction looks very much the same as would be expected for hadrons except for the very much smaller cross section, of course.

We now discuss the forward scattering in more detail including spin effects. The forward amplitude may be written

$$f(\nu) = f_1(\nu)\,\vec{e}_f\cdot\vec{e}_i + i\,f_2(\nu)\vec{\sigma}\cdot(\vec{e}_f \times \vec{e}_i) \tag{47.1}$$

as a spin matrix operating between spin states of the proton in the lab. system. The various measured quantities are expressed in terms of f_1 and f_2 as follows:

The total cross section is the imaginary part of the diagonal (in spin) scattering

$$\text{Im } f_1(\nu) = \frac{\nu}{4\pi} \sigma_{\gamma N}^{tot} \tag{47.2}$$

This is known. The forward differential cross section for unpolarized forward scattering is

$$\left(\frac{d\sigma}{dt}\right)_o = \frac{\pi}{\nu^2} (|f_1|^2 + |f_2|^2) = \frac{1}{16\pi} \sigma_{tot}^2 + \frac{\pi}{\nu^2} |\text{Re } f_1|^2 + \frac{\pi}{\nu^2} |f_2|^2 \tag{47.3}$$

The real part of f_1 can be obtained from the imaginary part by a dispersion relation, (Eq.(40.6a) for $q^2 = 0$) where we use the fact that $f_1(o) = -e^2/M$.

$$\text{Re } f_1(\nu) = -\frac{e^2}{M} + \frac{\nu^2}{2\pi^2} \int \frac{\sigma^{tot}(\nu')\, \nu' d\nu'}{\nu'^2(\nu'^2 - \nu^2)} \tag{47.4}$$

This has been evaluated (see Damashek and Gilman, Phys. Rev. D1 1319 (1970) or Buschhorn et al. Phys. Lett. 33B 241 (1970) and the sum of the first two terms of (47.3) is compared to the experimental $(d\sigma/dt)_o$ to see how big the last term is. They agree within errors for ν from 2.5 to 17 GeV so the $\pi|f_2|^2/\nu$ contribution is less than 10% over the entire range. (The contribution is greatest from the first term in (47.3) above about 5 GeV. The second is 15% at 2 GeV and falls away at higher ν).

We also know, for small ν, as $\nu \to 0$, $f_2(\nu) = -\frac{e^2}{2M} \mu_A^2 \nu$ where μ_A is the anomalous part of the magnetic moment of the nucleon in nuclear magnetons.

At finite angles the asymmetry parameter has been measured

$$\Sigma = \frac{\sigma_\perp - \sigma_\parallel}{\sigma_\perp + \sigma_\parallel} \tag{47.5}$$

where $\sigma_\perp$ and $\sigma_\parallel$ are differential cross sections at fixed t for incoming photons polarized perpendicular or parallel to the plane of collision. For $t = 0$ Σ must be zero, of course; but within the limits of experimental error (±10% for $-t < .2$, ±20% up to $-t = .6$) it is zero up to $t = -0.6$. (The average of Σ for t = .1 to .7 is .02 ±.06.)

We have discussed the size of $d\sigma/dt$ for γp in relation to (VDM) ρp cross section (see lecture 20), it is twice larger than VDM expects. The asymmetry produces no problem for the corresponding asymmetry in the ρ case, it is also very small. This is not unexpected, s channel helicity conservation also expects the same result. The question is: With incident light in the z direction with x polarization do more photons scatter at a small angle $\Theta \approx Q$ (transverse)/ν in the direction x or in the direction y? From the point of view of diffraction, currents generated by the incident wave must be adequate to produce the correct forward scattered wave to interfere with the incident wave to account for the loss of intensity of this wave represented by the total cross section. These currents are obviously limited to the spatial region of the proton, and so they produce scattered waves in other directions, the usual e^{At} of diffraction from the proton, just as in hadronic collisions. But these currents must make pure x polarization at least in the forward direction, they are x directed currents. In other directions at small angles we have the same intensity for x and y deflection except for a $\cos\Theta_{lab.} \approx 1 - \Theta^2_{lab}$ projection for x deflection, therefore $\Sigma \approx (1 - \cos\Theta_{lab.})/(1 + \cos\Theta_{lab.}) \approx \Theta^2_{lab.}/2 \approx -t/2\nu^2 \approx + .03$ for $-t = 0.6$, $\nu \approx 3.5$ where the data is taken. Thus we expect small Σ, if any, close enough to zero to not be in disagreement with experiment within its errors.

To summarize, the Compton scattering as a function of t above 2 GeV shows no surprises other than what we can expect from diffraction from the known total photon absorption cross section.

Below 2 GeV, therefore in the resonance region, there is no data. But it should be possible to make a pretty good theory of angular distribution and energy variation by considering a succession of s-channel resonances (many of the matrix elements of which are known from the study of $\gamma p \to \pi p$ in the same energy region, unknown ones may be guessed from the quark model). There is also a computable neutral pion exchange term

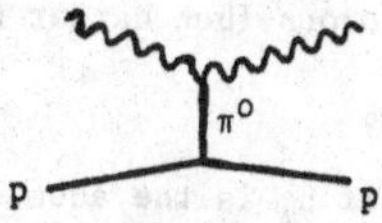

The coupling of two photons to a neutral pion is known from the $\pi^0 \to 2\gamma$ decay. All these calculations can be checked and controlled by fitting the calculated imaginary part of the scattering to the nicely measured total cross section $\sigma^{tot}_{\gamma N}$ which shows the expected resonance bumps in this region of energy.

<u>Compton effect for very small Q, ν</u>

Scattering of very low Q, ν is like the scattering of radio waves, or (if $q^2 \neq 0$) it depends on the reaction of a particle to nearly constant electric and magnetic fields. This is, of course, given by two experimental constants, (obtained by measurements in such fields) the charge and the magnetic moment (restrict to spin 1/2 case). Therefore we expect the Compton effect for low enough ν, Q to be given entirely in terms of these constants. The particles should act exactly as if they were point particles. One can compute the effect semiclassically or from the non-relativistic approximation to the Schroedinger equation with spin (Pauli equation) or again by diagrams assuming a pure particle with no internal excited states. Such a term is called a Born term. Thus we write $T_{\mu\nu}$ (not averaged over spin directions of the proton) as a sum from the Born term and the rest from other diagrams

$$T_{\mu\nu} = T_{\mu\nu}^{B} + T_{\mu\nu}^{R} \qquad (47.6)$$

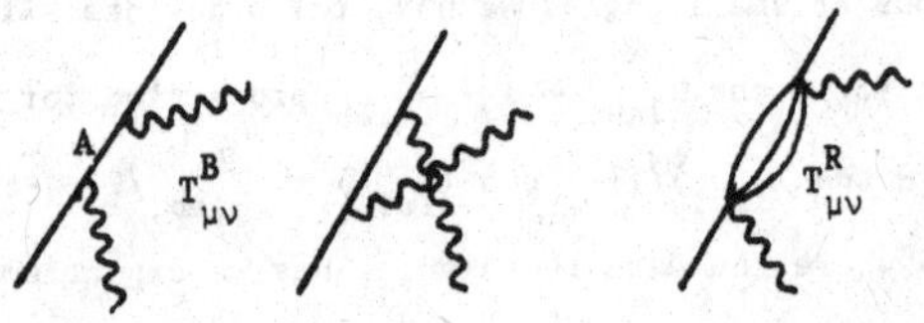

The reason T^B dominates at low ν is that it has an energy denominator due to the intermediate state A of size $(M_p + \nu - E_A)$ so if A is also a proton E_A is M_p (as $Q \to 0$) and we have a leading $1/\nu$ factor which doesn't appear in the remaining terms T^R.

As you expect the matrix element of the charge density $J_t = \rho$ at $Q = 0$ is the total charge and is diagonal thus $\langle x|\rho(Q)|p\rangle$ = order Q^2 if x is not the proton state. To show that the off diagonal matrix elements of other components of J_μ also go to zero we look at $q_\mu \langle x|J_\mu|p\rangle = 0$ (charge conservation); so $\nu \langle x|\rho|p\rangle = Q.\langle x|\vec{J}(o)|p\rangle$, hence if ν, Q go to zero together we see that $\vec{J}$ matrix elements go to zero. A more rigorous (but harder to interpret) argument is given below.

We now compute the limit of T^B as $\nu, Q \to 0$. If μ_A is the anomalous moment in nuclear magnetons, the coupling of a photon is $\gamma_\mu + \frac{\mu_A}{2M}(\gamma_\mu \not{q} - \not{q}\gamma_\mu)$ so

$$T_{\mu\nu}^{B} = \left(\gamma_\mu + \frac{\mu_A}{2M}(\gamma_\mu \not{q} - \not{q}\gamma_\mu)\right) \frac{\not{p} + \not{q} + M}{2p\cdot q + q^2} \left(\gamma_\nu + \frac{\mu_A}{2M}(\gamma_\nu \not{q} - \not{q}\gamma_\nu)\right) +$$

$$+ \left(\gamma_\nu + \frac{\mu_A}{2M}(\gamma_\nu \not{q} - \not{q}\gamma_\nu)\right) \frac{\not{p} - \not{q} + M}{-2p\cdot q + q^2} \left(\gamma_\mu + \frac{\mu_A}{2M}(\gamma_\mu \not{q} - \not{q}\gamma_\mu)\right) \qquad (47.7)$$

For small q and $q^2 = 0$ this is easily worked out to be

$$- \frac{e^2}{M} (e_f \cdot e_i) - i \frac{e^2}{2M^2} \mu_A \nu (\sigma \cdot \vec{e}_i \times \vec{e}_f)$$

while the contribution from T^R starts as ν^2 and thus we have as $\nu \to 0$

$$f_1(o) = - e^2/M$$

$$f_2(\nu)/\nu = f_2'(o) = - e^2 \mu_A^2/2M^2 \tag{47.8}$$

To show how T^R is small more formally (at least for the f_1 term) note that $T_{\mu\nu}$ total as well as T^B each separately satisfy the gauge condition $q_\mu T_{\mu\nu} = 0$, hence we must have $q_\mu T_{\mu\nu}{}^R = 0$. Now we can write $T_{\mu\nu}{}^R$ (for the symmetric spin averaged case at least) as a power series $T_{\mu\nu} = a p_\mu p_\nu + b(p_\mu q_\nu + p_\nu q_\mu) + c\,\delta_{\mu\nu}$ + order $q_\mu{}^2$. There must be no poles like $1/p\cdot q$ in a,b,c (unlike $T_{\mu\nu}{}^{total}$ which has such poles coming from T^B). Now $q_\mu T_{\mu\nu}{}^R = 0$ requires

$$(p\cdot q)P_\mu a + (p\cdot q)q_\mu b + q^2 p_\mu b + q_\mu c + \text{order } q^3 = 0$$

We cannot solve this by $a = -bq^2/p\cdot q$ because no $1/p.q$ terms are allowed. Clearly $c = - p\cdot qb$, $b = - p\cdot qa/q^2$ with $a = \alpha q^2$ are the only possibilities, therefore $a \sim \alpha q^2$, $b \sim - \alpha(p\cdot q)$, $c \sim \alpha(p\cdot q)^2$ and the term starts out as second order in q. The unsymmetric non-spin summed $T_{\mu\nu}{}^R$ can also be shown to be of the same order and therefore $T_{\mu\nu}{}^{Compton} = T_{\mu\nu}{}^B$ to order 1 and ν. (For a complete discussion see Low, Phys. Rev. 96 1428 (1954) and Gell-Mann, Phys. Rev. 96 1433 (1954).)

<u>Forward Compton scattering from non-relativistic Schroedinger equation</u>

The equation is (with first order relativistic corrections)

$$H\psi = \left[\frac{1}{2M} (\vec{p} - e\vec{A})\cdot(\vec{p} - e\vec{A}) - \frac{1}{8M^3} (\vec{p}\cdot\vec{p})^2 - \frac{e}{2M} (1 + \mu_A)\vec{\sigma}\cdot\vec{B} + \right.$$

$$\left. + \frac{e^2}{8M^2} (1 + 2\mu_A)(\nabla\cdot\vec{E} + 2\vec{\sigma}\cdot(\vec{p} - e\vec{A})\times\vec{E})\right] \psi = E\psi \tag{47.9}$$

The incident amplitude is $\vec{A} = \vec{e}_i$, $\vec{E} = i\nu\vec{e}_i$ and $\vec{B} = i\,\vec{k} \times \vec{e}_i$. In the laboratory $\vec{p}$ is zero, hence the leading term comes from the $\vec{A}\cdot\vec{A}$ term and is $-(e^2/2M)\vec{e}_i\cdot\vec{e}_f$ contributing to f_1. Next we have the term $\vec{\sigma}\cdot\vec{B}$ operating in second order, two diagrams with energy denominators $-\nu$ and $+\nu$

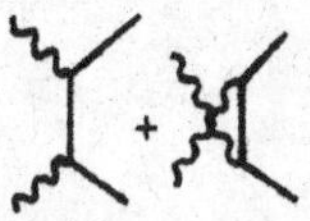

$$= \left[\frac{(\vec{\sigma}\cdot\vec{k} \times \vec{e}_f)(\vec{\sigma}\cdot\vec{k} \times \vec{e}_i)}{-\nu} + \frac{(\vec{\sigma}\cdot\vec{k} \times \vec{e}_i)(\vec{\sigma}\cdot\vec{k} \times \vec{e}_f)}{\nu}\right]\left(\frac{e}{2M}\right)^2 (1+\mu_A)^2$$

This is $- \frac{1}{\nu} i\sigma \cdot ((\vec{k} \times \vec{e}_f)\times(\vec{k} \times \vec{e}_i)) = -i\sigma \cdot (\vec{e}_i \times \vec{e}_f)\nu$ so we get

$$-i \frac{e^2}{2M^2} (1 + \mu_A)^2 \nu \, \vec{\sigma}\cdot(\vec{e}_i \times \vec{e}_f) \text{ contributing to } f_2$$

Finally the last term in (47.9) with the term $\vec{\sigma}\cdot\vec{A} \times \vec{E}$ gives

$$\frac{e^2}{4M^2} (1 + 2\mu_A) \left[\vec{\sigma}\cdot(\vec{e}_i \times i\nu\vec{e}_f) + \vec{\sigma}\cdot(\vec{e}_f \times (-i\nu\vec{e}_i))\right]$$

which combines with the previous term to change the $(1 + \mu_A)^2$ to μ_A^2 we therefore have

$$\text{Amp.} = -\frac{e^2}{2M} \vec{e}_i \cdot \vec{e}_f - i \frac{e^2}{2M^2} \mu_A^2 \nu \, \vec{\sigma}\cdot(\vec{e}_f \times \vec{e}_i) \quad .$$

Other Two-Current Effects

Lecture 48

<u>Other quantities involving $T_{\mu\nu}$</u>

Another experimental quantity that involves our function $T_{\mu\nu} = \langle p|\{J_\mu J_\nu\}_T|p\rangle$ (not averaged over the spin of the proton, it involves the antisymmetric part of $T_{\mu\nu}$) is the hyperfine splitting energy in hydrogen responsible for the 1420-megacycle line. It is the difference in energy in the ground s state of atomic hydrogen depending on whether the spins of the electron and proton are parallel or anti-parallel. In non-relativistic approximation it depends on the probability that the electron is on top of the proton $|\psi(o)|^2$ in the ground state wave function. Relativistically we can write (R_∞ = Rydberg, μ_p, μ_e magnetic moments of p and e, μ_o = Bohr magneton)

$$\delta E = \frac{32\pi\alpha^2}{3} R_\infty \frac{\mu_p \mu_e}{\mu_o^2} \left(1 + \frac{m}{M}\right)^{-3} \left(1 + \frac{3}{2}\alpha^2\right) \mathcal{E}\, \mathcal{R}\, \mathcal{P} \qquad (48.1)$$

The factor $(1 + m/M)^{-3}$ comes from reduced mass corrections to the Schroedinger equation in getting $|\psi(o)|^2$; $1 + 3\alpha^2/2$ is a modification due to the Dirac equation. The other factors $\mathcal{E}$, $\mathcal{R}$, $\mathcal{P}$ are all near one and are due to higher order quantum electrodynamic corrections. They have been separated into three factors for convenience of discussion. $\mathcal{E}$ comes from QED modifications of the motion of the electron, diagrams like A below.

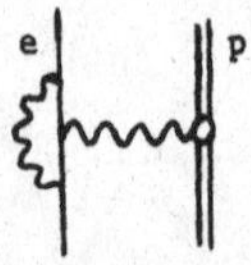

A

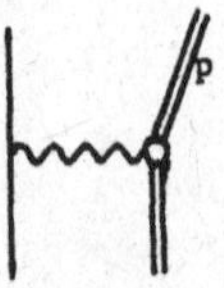

B

$\mathcal{R}$ comes from the proton recoil diagrams of type B in which the fact that the proton is not a point charge is included by using measured form factors. $\mathcal{P}$ comes from two photon exchange terms of the form C, D.

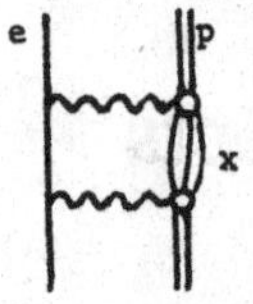

C

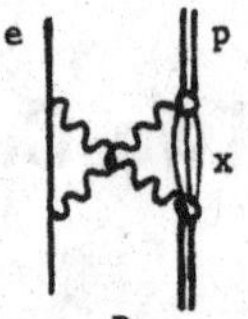

D

All these factors except $\mathcal{P}$ which depends on as yet unknown properties of the proton have been calculated to very high accuracy - further δE is measured to absurdly high accuracy. The constants like α, μ_p etc. are now known well enough to determine $\mathcal{P}$ to about four parts per million. Theoretically, the deviation due to $\mathcal{P}$ is about this same order of magnitude. As long as $\mathcal{P}$ remains uncertain we cannot use these accurate measurements and calculations to improve our knowledge of the constants - or to put the problem the other way if more accurate values of the other constants become available that will tell us $\mathcal{P}$, an electromagnetic property of the proton - and we would be challenged to calculate it, write $\mathcal{P} = 1 + \Delta$. Let us see what is involved.

Obviously the proton coupling is to two currents. We separate the cases that the photons are of low energy and momentum from the cases where they are high. Where they are low, binding of the electron in initial and intermediate states etc. must be considered - but here for low ν, Q, the proton acts, as we have seen, like a particle of charge and magnetic moment and thus we can do this part of the calculation. For definiteness we do it putting in the experimental form factors once for each photon and integrate over all moments, call this Δ_1. Write $\Delta = \Delta_1 + \Delta_2$. Now for high virtual momentum electron binding, even the electron mass can be neglected and we can imagine the electron and proton to be free and at rest before and after the scattering - thus our two current operator $\langle p|J_\mu J_\nu|p\rangle$ for proton in and out of same momentum (rest) is all that is involved. Naturally states x other than pure proton Born states are involved. Of course we have already counted some high momentum contributions in Δ_1 and we shall have to subtract them, thus we shall have not just $T_{\mu\nu}$ but rather $T_{\mu\nu} - T_{\mu\nu}^{\text{elastic}}$ where $T_{\mu\nu}^{\text{elastic}}$ is calculated from the Born term alone with form factors. Call this $T_{\mu\nu}{}' = T_{\mu\nu} - T_{\mu\nu}^{\text{elastic}}$. Then Δ_2 is proportional to diagrams C, D with electron at rest initially and finally. If $T_{\mu\nu}{}'$ is a γ matrix on the proton spin it depends on the spin flip amplitude for the proton – thus on the antisymmetrical part of $T_{\mu\nu}{}'$ (proton at rest).

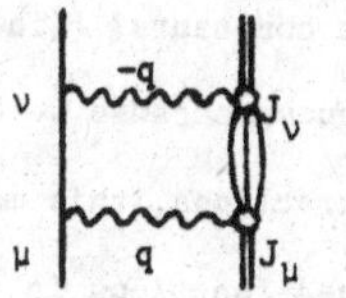

$$\Delta_2 \sim \int \frac{d^4q}{(q^2)^2}\, \mathrm{Tr}\,(C_{\mu\nu}\, \tfrac{1}{2}(1+\gamma_t)\gamma_z\gamma_5)\,\mathrm{Tr}\,(T_{\mu\nu}{}'\, \tfrac{1}{2}(1+\gamma_t)\gamma_z\gamma_5) \qquad (48.2)$$

The factor $1/(q^2)^2$ is for the two photon propagators. The first trace is the Compton scattering by the electron

$$C_{\mu\nu} = \gamma_\nu \frac{1}{\not p + \not q - m}\gamma_\mu + \gamma_\mu \frac{1}{\not p - \not q - m}\gamma_\nu \approx (\gamma_\nu \not q \gamma_\nu - \gamma_\nu \not q \gamma_\nu)/q^2$$

neglecting the momentum p and mass of the electron. The second factor is the coupling of two photons to the proton, which we do not know but which we are discussing; we need its antisymmetric part. In Lecture 33 we wrote its imaginary part $W_{\mu\nu}$ in the form

$$W_{\mu\nu} = 4\,P_\mu P_\nu W_2 - 4M^2\delta_{\mu\nu}W_1 + 4Mi\epsilon_{\mu\nu\lambda\sigma}\, q^\lambda\, [M^2 s^\sigma G_1 + (P\cdot q\, s^\sigma - s\cdot q\, P^\sigma)G_2]$$

where G_1 and G_2 are functions of q^2 and ν defining the imaginary part. Let the complete scattering functions of which G_1 and G_2 are the imaginary part be called S_1 and S_2. That is, we write a form for $T_{\mu\nu}$ just like the above for $W_{\mu\nu}$ except S_1, S_2 replace G_1, G_2; and G_1, G_2 are Im S_1, Im S_2. Then by substitution we can express Δ_2 directly in terms of S_1, S_2. One gets (see Drell and Sullivan, Phys. Rev. <u>154</u> 1477 (1967) and C.N. Iddings, Phys. Rev. <u>138B</u> 446 (1965)).

$$\Delta_2 \propto \int \frac{d^4q}{(q^2)^3}\left[(2q^2 - \nu^2)\, S_1\,(q^2,\nu) + \frac{3q^2\nu^2}{M^2}\, S_2\,(q^2,\nu)\right]$$

Now we can do many things with this to try to estimate it or compare it to things measurable in principle in ways entirely analogous to our discussion of electromagnetic self energy. For an example, we can use Cottingham's idea of rotating the contour on the dq_0 integral from the real to the imaginary axes, thus the integrals depend only on S_1, S_2 in the negative q^2 region. Finally we can express these complete S_1, S_2 in terms of G_1, G_2 (their imaginary parts) by a dispersion relation, and thus express Δ_2 in terms of G_1, G_2.

Two questions come up:

First, are these unknown functions, like $T_1(o,q^2)$ necessary in the dispersion relation or are they as for T_2 without constants? (The answer to this question is almost certainly known.) We can guess because we know the asymptotic scaling behavior of G_1, G_2. If there are subtractions, this method is frustrated.

Second, supposing there are no undetermined functions in the dispersion relation and Δ_2 can be entirely expressed in terms of G_1, G_2 in the experimental region; what can we do until G_1, G_2 is directly measured? It becomes a research problem to guess as completely as possible to see in what ranges of $-q^2$, ν the Δ_2 is most sensitive and use whatever models or ideas are most reliable there. We could try to incorporate all that is known of low energy theorems and integrals (like $\int g_2 dx = 0$) to limit the possibilities. At worst, certain limits of uncertainty can be established since, for example, 2g, which is the difference of up and down spin partons $h_+(x) - h_-(x)$ (see Lecture 33) cannot exceed the sum $h_+(x) + h_-(x)$ which is f(x) and is measured. Generally positivity of the trace of Im $T_{\mu\nu}$ on any diagonal state limit the size of G_1, G_2 in terms of W_1, W_2 in some way. This problem should be pursued

And if the answer to the first question is yes and there are unknown functions brought in in the fixed q^2 dispersion relations? Then Cottingham's scheme does not work and we shall have to use other methods of analysis such as the fixed Q dispersion relations to obtain expressions on which we can apply our partial physical understanding of photon hadron interactions - partons, scaling, quark model, etc., - in conjunction with as many sum rules etc., that we know (as well as possibly even the numerical value of the n-p mass differences) to guide us as much as possible to calculate this quantity Δ_2 as accurately as possible, and with an honest estimate of the possible theoretical limits of uncertainty.

Lecture 49

Other two-current effects

I should just like to add a few miscellaneous remarks of situations in which the double operation of two currents is involved, namely, in the disintegration of pseudoscalar mesons. They all present interesting questions for study; this is meant merely as an introduction. Of course matrix elements of one J_μ are involved in single photon decays like $\omega \to \pi\gamma$ and we have already discussed them.

Two J's are obviously involved in two photon decays like $\pi^o \to 2\gamma$ or $\eta^o \to 2\gamma$. An honest calculation of either one of these would be very interesting (one in which the validity of the assumptions is backed up by considerably more than the mere fact that the answer agrees with this one experiment). What does SU_3 say? Use the quark model $\pi^o = \frac{1}{\sqrt{2}}(u\bar{u} - d\bar{d})$, $\eta^o = \frac{1}{\sqrt{6}}(u\bar{u} + d\bar{d} - 2s\bar{s})$ to get the ratio of amplitudes amp π^o/amp $\eta^o = \frac{1}{\sqrt{2}}(\frac{4}{9} - \frac{1}{9})/\frac{1}{\sqrt{6}}(\frac{4}{9} + \frac{1}{9} - 2\frac{1}{9}) = \sqrt{3}$. It says that the amplitude for π is $\sqrt{3}$ that for η, so the rate should be three times as high. Experimentally the widths are $\Gamma(\pi^o \to 2\gamma) = 7.2 \pm 1.0$ eV, and the partial width $\eta \to 2\gamma$ is $\Gamma(\eta \to 2\gamma) = 1.0 \pm .3$ KeV. The ratio is 1/140 instead of 3!

Of course, the reason for the abject failure is the very large mass difference of π and η ($m_\pi^2/m_\eta^2 = 1/15$) and we must be much more careful of it. This is where SU_3 is indefinite - and no universal way of doing this is known. First there is phase space: the general formula for disintegration of an object of mass m at rest into two particles each of whose momentum is Q is

$$\Gamma = \frac{Q}{8\pi m^2}|M|^2$$

where M is the relativistic matrix element. In our case (2γ) $Q = m/2$ so phase space (i.e. if we assume M is given by SU_3) works against the η and the discrepancy is another factor of m_η/m_π worse (i.e. $M_\pi^2/M_\eta^2 = 1/540$ instead of 3/1).

More sensibly we should write M in its simplest relativistic form which for a pseudoscalar meson disintegrating into 2γ's of polarization e_1, e_2, momenta k_1,k_2 is $M = a\,\varepsilon_{\mu\nu\sigma\rho}\, e_{1\mu}\, e_{2\nu}\, k_{1\sigma}\, k_{2\rho}$.

Now the guess would be that a is determined by SU_3. This means M goes as Q^2 or m^2 $\therefore$ Rate$_\pi$/Rate$_\eta = \frac{m_\pi^3}{m_\eta^2}$, $a_\pi^2/a_\eta^2 = \frac{1}{140}$. This at least moves in the right direction, it gives

$$a_\pi^2 / a_\eta^2 = 0.4 \quad \text{instead of } 3.0 \,.$$

This is not bad but (a) what of the remaining discrepancy and (b) why should a be given by SU_3 and not say a/m or a/m^2? Evidently we have been doing too much comparison to experiment and too little thinking. Can we reason out from other things we know just how the π rate and η rate should be compared, or how either might be calculated or estimated absolutely?

With regard to the mechanism of the decay ever since it was suggested by Gell-Mann, Sharp and Wagner, Phys. Rev. Letters 8 261 (1962) it is supposed to be dominated by the diagram going through an intermediate ρ or ω meson like

This connects it to the $\eta\rho\gamma$ coupling constant, or through SU_3 to the pseudoscalar vector photon coupling constants in general; determined directly for example by the $\omega \to \pi\gamma$ rate.

Again $\eta \to \pi\pi\gamma$ is interpreted in the same way as

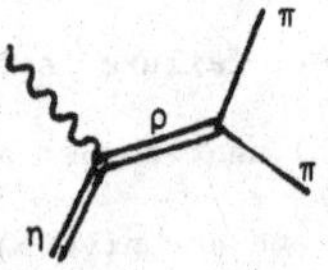

The $\rho\pi\pi$ coupling being known one can compare this with $\eta \to \gamma\gamma$ and predict the ratio $\Gamma(\eta \to \pi\pi\gamma)/\Gamma(\eta \to \gamma\gamma)$ with good success. Gell-Mann et al. got .25, experiment is .12 (Gormley, Phys. Rev. 2D 501 (1970)). (For two calculations corresponding to different choices of how the coupling constants symmetry rules depend on the masses of the states see Brown, Muncek and Singler, Phys. Rev. Letters 21 707 (1968) and Chan, Clavelli and Torgerson, Phys. Rev. 185 1754 (1969). The choice made in the latter paper fits very nice.) This is interesting for at first sight the numerator is order α and denominator is order α^2 so the order should be 137 but here many numerical constants accumulate to overwhelm this factor and make the ratio nearly 1000 times smaller. For this picture there is a factor in the matrix element for $\eta \to \pi\pi\gamma$ like $1/(m_\rho^2 - m_{\pi\pi}^2)$ where $m_{\pi\pi}^2$ is the invariant mass squared $(p_+ + p_-)^2$ of the four-vector sum of the two pions. This distorts the spectrum away from the simplest form toward larger probabilities for larger $m_{\pi\pi}^2$. The experiments (Gormley) are so accurate to see this effect, even quantitatively, so there can be no doubt of the mechanism in this case. It is likely therefore that no deep mysteries lie in the 2γ disintegrations either.

However the $\eta \to 3\pi$ does present a challenge. The G parity of the π's is -, of the η is + so the disintegration is not allowed, strongly. Consider $\eta \to \pi^0\pi^+\pi^-$, charge conjugation (which we think is surely satisfied for decays at this rate) (η and π^0 are charge conjugation + because they can go into 2γ) requires that the π^+, π^- must be symmetric, they must be in an I = 0 or 2 state. This added to the I = 1 of the third meson yields only total I = 1, 1, 2, 3 as possibilities, there is no I = 0. Therefore the decay cannot occur except by violation of isotopic spin. Isospin is, of course, violated by electrodynamics so an electrodynamic virtual effect (order α in matrix element) is involved and the rate is of order α^2 so $\eta \to 2\gamma$ can compete with it rather than being completely swamped by what it would be if $\eta \to 3\pi$ were a strong interaction. (The data on η decays is

	Branching ratio %
$\eta \to \pi^+\pi^-\pi^0$	23.1 ± 1.1
$\to 3\pi^0$	30.3 ± 1.1
$\to \pi^+\pi^-\gamma$	4.7 ± 0.2
$\to \gamma\gamma$	38.6 ± 1
$\to \pi^0\gamma\gamma$	3.3 ± 1

$\Gamma^{total}(\eta) = 2.70 \pm .67$ KeV)

The change in I spin can be $\Delta I = 0, 1, 2$ but only the $\Delta I = 1$ part violates G parity. Hence the final state of three pions must have I = 1. From this we can get an estimate of the ratio $\Gamma(\eta \to 3\pi^0)/\Gamma(\eta \to \pi^+\pi^-\pi^0)$. Combining three states of isospin one we can get for the total I spin:

symmetric	3, 1
skew symmetric	2, 2, 1, 1
antisymmetric	0

If we suppose the 3π are in their lowest space wave, s wave and symmetric, since they are Bose we will have I spin 3 or 1; EM permits only the 1 (but by proper use of space states the skew-symmetric I = 1 states could come in).

For this state $\Gamma(\eta \to 3\pi^0)/\Gamma(\eta \to \pi^+\pi^-\pi^0) = 3/2$. If there is some skew symmetric (momentum dependent) space state the other I = 1 can come in reducing the 3/2; since it comes in with higher angular momentum it is probably smaller so the ratio may be fairly close to 3/2. It is 1.3 experimentally.

How do we calculate the rate $\eta \to 3\pi$? Where does the intermediate photon act; can we guess which intermediate states are most likely to be important? A quantitave estimate for this rate is a problem that no discussion of the effects of virtual self energy photon action can fail to mention.

Hypotheses in the Parton Model

Lecture 50

Hypotheses in the Parton Model

We should now like to discuss what we can say about what the products X would look like in deep inelastic scattering $e + p \to e + X$. There are some measurements for certain definite final states X for small energies and low q^2 of the virtual photon (see Berkelman, 1971 Cornell conference). Most of these can be understood from direct extensions of our theories for photon ($q^2 = 0$) reactions. We have already discussed pion production from virtual pion exchange as yielding information on the pion form factor. In addition ρ production has been studied from virtual photons, with no surprises - VDM gives a fair account - see our theory discussed in Lectures 16 to 21 where we merely have to replace k^2_{out} by q^2 in the equations there yielding a factor $m_\rho^2/(m_\rho^2-q^2)$ (q^2 is negative) relative to the $q^2 = 0$ case.

It is necessary also to make an assumption of how the longitudinally polarized photon (possible when $q^2 \neq 0$) couples in relation to the longitudinally polarized ρ — it is assumed that these amplitudes are related by a factor $(q^2/m_\rho^2)\dfrac{m_\rho^2}{m_\rho^2 - q^2}$ the extra factor q^2/m_ρ^2 is an ansatz made because gauge invariance requires that J longitudinal vanishes as $q^2 \to 0$. This may be valid for not too large q^2, but of course if q^2 becomes really very large the assumptions about small $\rho \to \rho^*$ production may begin to fail.

What happens at large q^2, and how it ties on to small q^2, in every case for very large ν will be our present concern. We shall have to be guided by theory and I will take this opportunity to review the parton model and some assumptions that can be made about it. We shall list assumptions that we can make - without today being sure of which are right and which wrong - just to see what their consequences are with the hope that experiment may later make the selection (e.g. are charged partons quarks?). Therefore in our list some assumptions will (perhaps) be inconsistent with others. The assumptions will come from a mixture of theoretical guesses and known experimental facts - so one might be warned that if a particular assumption neatly explains some experimental fact it may not really be a significant confirmation for the assumption might have been made with that fact in mind. Finally, little effort will be made to derive one assumption from another - they are certainly not independent. This will therefore unfortunately not be a mathematical and sound system, but rather a lengthy "intuitive" or physical discussion.

For a good discussion along the same lines see J. Bjorken's paper in the 1971 Cornell Conference.

<u>General Framework</u>

We suppose as in field theory, a wave function for a state can be given by giving the amplitude for finding various numbers of field quanta, or partons of various momenta. In particular we discuss the wave functions of single particles (sometimes two particles in collision) with extremely large momentum P in the z direction ($P \to \infty$). The wave function is being described in Fock space giving for a state the amplitude

$$\begin{aligned}\psi = \;&\psi_o\\&\psi_1(p_1)\\&\psi_2(p_1, p_2)\\&\cdot\\&\cdot\\&\text{etc.}\end{aligned}$$

where ψ_o is the amplitude to find no partons (usually zero); $\psi_1(p_1)$ is the amplitude there is one parton (of such and such a type, an index we are omitting) which has momentum p_1; ψ_2 is the amplitude there are two partons of momenta p_1, p_2 etc. This can be written in other ways. For example let |VAC> represent the vacuum state and a^*_p the operator creating a parton of momentum p. Then we can write the wave function state $|\psi\rangle = F^* |VAC\rangle$ where

$$F^* = \psi_0 \cdot 1 + \sum_{p_1} \psi_1(p_1)\, a^*_{p_1} + \frac{1}{2}\sum_{p_1 p_2} a^*_{p_1} a^*_{p_2}\, \psi_2(p_1,p_2) + \ldots$$

is some function of the creation operators.

Then we make the following assumptions:

<u>A1</u>. The amplitude to find a large $P_\perp$ on any parton falls rapidly with $P_\perp$ such that effectively we can in first approximation <u>consider all $P_\perp$ are finite</u> (as $P \to \infty$).

<u>A2</u>. The "wave function" for longitudinal momentum of order P, i.e. <u>$p_L = xP$ depends only on x.</u>

This requires some complicating remarks to make its definition clear for there is, as $P \to \infty$, an ever increasing contribution to ψ for small x. More precisely consider the density matrix. Let $\psi_m(p_1 p_2 \ldots p_n)$ be the amplitude to find n partons of momenta p_1 to p_n. Then for example the density for one at k is

$$\sum_n \int |\psi_n(p_1,p_2,\ldots\ldots p_n)|^2 \sum_i \delta(k-p_i) \prod_i d^3p_i = \rho(k)$$

That this depends only on $k_\perp$ and $x = k_L/p$ when x is finite as $P \to \infty$ is the assumption we want to make - with all its generalizations. E.g. the one particle density matrix (like $\sum_n \int \psi^*_n(p_1,p_2,..k..p_n)\, \psi_n\, (p_1 p_2 ..k' \ldots p_n)$ $dp_1 \ldots dp_n = \rho(k,k')$ depends only on $k_\perp, k'_\perp, k_L/P, k'_L/P$ etc., when x is finite. Again the two particle density (the expectation of $\sum_{ij} \delta(p_i-k_1)\delta(p_j-k_2)$) behaves likewise etc. It is almost but not quite the same as saying the wave function in $p_{\perp i}$, $x_i = P_{Li}/P$, has a definite limit. The wave function is a function of all of the momenta including those of finite momentum (which we call wees). The scaling doesn't work for those - in fact the mean number of particles rises with P so the wave function ψ_n for any fixed n falls with P (like a power of P) but the "relative wave function" e.g., a ratio in which only one particle (or a finite number) is moved depends for finite x, $p_\perp$ only on x, $p_\perp$.

<u>A3</u>. In the wave function the amplitude to find <u>finite longitudinal</u>

momentum particles remains finite as $P \to \infty$. That is to say again that the density matrix, e.g. the density for finding particles with finite (e.g. < +4 GeV) values of p_L have a definite limiting behavior as $P \to \infty$, and the expected number of such wee partons is finite.

Lecture 51

Hypotheses in the Parton Model (continued)

A4. To have continuity between A3 and A2 the mean number of partons of a given type for small x goes as dx/x and the wees go as dp_z/p_z as $P \to \infty$. The number for p_z negative falls off rapidly so that for $p_z = xP$, finite x negative there are no partons; although for p_z negative and finite there are some (a fixed amount falling rapidly with negative p_z).

(As a trivial example of the kind of behavior envisaged in the wee region consider a wave function like $\exp(\Sigma C_k a_k^*)|0\rangle$ where a_k^* creates a particle of longitudinal momentum k and C_k varies as $C_k = \alpha/(\omega-k)\omega^{3/2}$ with $\omega = \sqrt{k^2+1}$ say, and α = constant.)

A5. The behavior of the wees is nearly (as $P \to \infty$ completely) independent of the distribution of the fast (finite x) partons. This again is complicated. If we stretch out the variable p_z by, for example, defining $y = \ln(\sqrt{p_z^2 + 1\ \text{GeV}^2} + p_z)$ so for finite p_z, y is finite; for finite x,y is $\ln 2P + \ln x$. We have particles at every y from finite to $\ln 2P$.

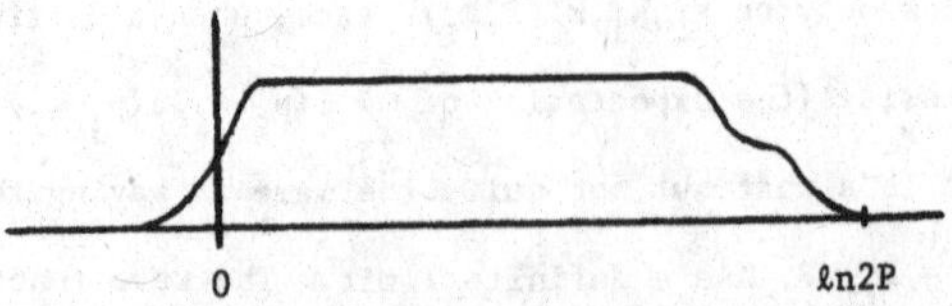

We have drawn a graph of the mean number of particles in dy. If we look near finite y, $P \to \infty$ we see the behavior of the wees; if we look at finite $\ln 2P - y$ (finite x) we see the behavior of the scaling fast ones; in between is a plateau with a finite density of partons so the mean number of partons rises as $\ln P$.

It is easy to understand the density but how do we understand the wave function? This gives the amplitude for every configuration, which is a set

of values of y for partons present

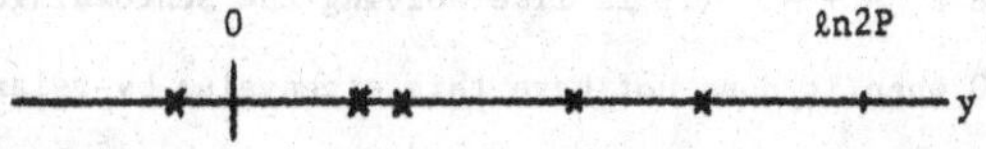

How does this amplitude vary with the position of the dots? It behaves like a wave function for a finite one dimensional drop of liquid of thickness $\ell n2P$. The amplitude is large only if particles are more or less everywhere, with a uniform density except near the surfaces at $\ell n2P$ and 0. The behavior at one surface is not strongly influenced by what the configuration is at the other surface - they are insulated from each other by the long (?, $\ell n2P$ is never really very large) intermediary plateau.

Mathematically we are discussing the solution of $H\psi = E\psi$ for a state of fixed P_z (but we do not know H, of course). If $\mathcal{P}_z$ is the momentum operator (e.g. like $\Sigma\, p_z\, a_k^* a_k$ etc.) we want $\mathcal{P}_z\, \psi = P\psi$. Consider then that ψ is an eigenvector of the operator $W = H - \mathcal{P}_z$, for a state of mass M. Now as $P \to \infty$ $(E = \sqrt{P^2+M^2} \approx P+M^2/2P)$ we can consider the limit

$$(2PW)\psi = M^2\psi \qquad (51.1)$$

so we are looking at eigenvectors with fixed eigenvalues of the operator PW as $P \to \infty$. We are assuming that it looks as if, as $P \to \infty$ the operator PW has a distinct limit, expressible in $x = p_L/P$ and $p_\perp$. This would be nice, but there is trouble on the small momentum end. The equation is like a cascade, large x generates smaller x through interaction terms (like turbulence equations, or cosmic ray showers). Smaller momenta pile up until new phenomena sets in to change the equations (like viscosity in turbulence, or ionization loss in cosmic rays) to finally determine the wee x (finite p_z) behavior. (The approximation in showing that PW depends only on x is wrong, for example we can no longer write $\sqrt{p_z^2+p_\perp^2+m^2} \approx p_z$.) But by that time the "shower is fully developed" and the behavior of the wees (except for normalization - total strength of wees) is independent of the way it started at finite x. (In the wee region interaction energies are comparable to kinetic energies.)

The behavior at this end is a solution of the equation

$$W\psi = 0 \qquad (51.2)$$

(Note the omission of P.) In general the operator W does not have a zero eigenvalue if all boundary conditions are used - but here we relax the boundary condition of finiteness as $p_z \to +\infty$. (It is like solving the Schroedinger equation $H\psi = E\psi$ for $E = 0$ when it does not have this eigenvalue by relaxing say the condition at $r \to \infty$ and thus studying open scattering states approximately to which the real large r behavior will have to be attached - here we must ultimately really attack the finite x solution of $2PW\psi = m^2\psi$.)

It can be shown that since equation (51.2) is invariant under a boost (a Lorentz transformation in the z direction by velocity v, call $f = \sqrt{(1+v)/(1-v)}$) in which all large p_z are multiplied by f, that ψ can be of the form $f^\beta\psi$ (so boosting does not change ψ, only the normalization). This means that for finite but small x the probability of one parton at x varies as $x^{2\beta}dx/x$. The lowest β solution we assume in A4 corresponds to $\beta = 0$ (from experiment, not theory). Other solutions exist for higher β and the general solution is a linear combination of these whose coefficients are determined by how they fit on to (51.1). I had hoped to get a field theory interpretation of Regge theory this way but I have not completed the analysis.

<u>A6. The distribution of the wees is the same for all hadrons</u>. This is a bold assumption partly guided by experiment. Among other things it means the wees are neutral to isotopic spin; the wees for the proton look just as they do for the neutron. The difference can only come from a higher β and hence falls in amplitude relatively as $P^{-\beta}$ $(\beta > 0)$. The assumption that the wees are SU_3 symmetric will <u>not</u> be made, (I think it leads to disagreement with experiment in the expected ratios of π's to K's in certain experiments) for we imagine interaction forces are effective in determining the wee distribution and such forces are not SU_3 invariant.

(It strikes me at this moment, that since the wees are determined by $W\psi = 0$, i.e., a state of zero mass squared, and that since pions have a small mass, the state of the wees may be approximately only pions (with kaons, of larger mass much reduced, hence large SU_3 breaking.) Known pion interactions (perhaps described by intermediary ρ mesons) might permit a solution of $W\psi = 0$ in terms appropriate pion base states. If you do this I would suggest it might be easier if you work at first with the symmetric (in$\pm p_z$) wee distribution corresponding to two fast hadrons colliding, rather than this one-sided, one-particle distribution.)

The assumption A6 is not completely obvious from field theory - for there might be some long range direct effect of the fast partons on the slow ones in principle. The choice is guided by experiment (which shows that the right-moving products of hadronic collisions depend only on the right-moving initial colliding particle and not on what it collided with. To see how this assumption is used see B1, and J. Benecke et al. Phys. Rev. 188 2159 (1969). The physical assumption B1 that we make later says specifically that there is no such long range effect.

A7. Continuity demands, since the wees are adjacent to the sea (plateau region) that we also have that the sea (e.g. mean numbers and correlations of partons) is the same for all hadrons.

A8. The probability there are no partons in a sufficiently large gap of rapidity Δy (see A5) goes as $e^{-\gamma\Delta y}$ where γ depends on the quantum numbers (angular momentum, isospin, strangeness) carried by the gap. E.g. suppose we have a proton state and we ask for certain partons a,b,c, for $y > y_1$ and others below y_2

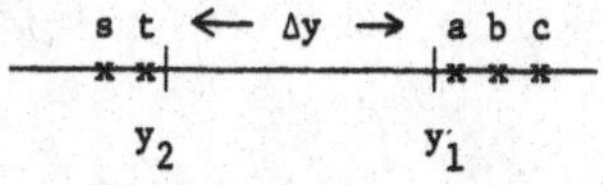

Perhaps their strangeness is + 1, the entire proton has S = 0 so there is a contribution S = - 1 across the gap. It is evident that this "quantum numbers carried across the gap" defined as the quantum numbers of the state minus those to the right of the gap (a,b,c) is just the sum of those to the left (s,t). This more complicated method of expression is an anticipation of the same idea for distributions when two hadrons are colliding. Then it is the quantum numbers of the hadron moving to the right minus the total quantum numbers of the partons to the right of the gap.

Assumption A8 is not stated clearly. We have to say how the gap Δy is changed. It is used in two cases: 1) Region a,b,c and s,t are both stated completely and the gap widens simply because P increases; hence $\Delta y \propto \ln 2P$ and the amplitude falls as $P^{-\gamma}$. This was used in analyzing the proton form factor in lecture 29, for example. 2) The gap is in a plateau. On one side or both there is a large stretch of plateau. Here a,b,c is fixed and s,t etc., is anything at all over a wide range of y up to the other boundary (order $\ln 2P$

away) Δy is kept fixed as $\ell n 2P$ rises. Probability goes as $(x_2/x_1)^{-\gamma}$ where x_1, x_2 are the x values at each end of the gap (used for asymptotic behavior of deep inelastic scattering near $x = 1$).

Hadron-Hadron Collisions at Extreme Energies

Lecture 52

Hadron-hadron collisions at extreme energies

Although our subject is photon-hadron interactions we shall review the assumptions made in describing hadron-hadron collisions A + B at extreme energies. We first leave out elastic scattering and diffraction dissociation and aim toward the large part of the cross section where several particles are emitted $A + B \rightarrow C + D + E + \text{---}$.

For a hard collision suppose the momenta of A, B are P_A, P_B respectively in the z direction - for example take center of mass $P_A = P_B = P$. (For a finite z-velocity v transformation from this $P_A = fP$, $P_B = \frac{1}{f} P$ with $f = \sqrt{(1-v)/(1+v)}$.) We only work with P_A, P_B or $P \rightarrow \infty$.

The asymptotically incoming wave function will be, of course, (i.e. before "interaction"), some kind of product wave function of A of momentum P_A to right and B of momentum P_B to left. Technical problems arise here. In field theory this cannot simply be a product of the wave functions of each particle that we have been describing for that is not unique (for example suppose A contains a Fermion parton at momentum p, and B also contains one of the same kind at the same momentum,but there cannot be two in the field as they are Fermions). Thus if A is represented as a creation operator F_A^* on the vacuum,

$\psi_A = F_A^*|VAC>$ which creates all our partons, and B by F_B^* we can define the incoming asymptotic wave function as

$$F_B^* F_A^* |VAC>$$

There is some trouble in the wee region where $F_B^* F_A^*$ do not commute (note that no creation operators for fast partons $p \sim xP_A$ appear both in F_A^* and F_B^* because A and B are moving in opposite directions). Actually this is only technical because we only want the state after the interaction. The problem would arise only if we were quantitatively calculating the interactions; but now we wish to talk about how the wave function looks after interaction hence say after "interaction plus correction for overlap in defining the initial state". The overlap affects only the wees, but we shall assume the interaction affects only the wees also. (By interaction we mean the effects of the fact that although $F_A^* |VAC>$ and $F_B^*|VAC>$ are both eigenfunctions of $H|\psi> = E|\psi>$, $F_A^* F_B^* |VAC>$ is not.)

Following we state the assumptions we shall make regarding the interacting wavefunction.

B1. Partons interact only if their relative four-momentum is finite, assuming they have some finite mass of order 1 GeV. This is equivalent to the statement, if $y = \ln(\sqrt{p_z^2+1} + p_z)$ (in GeV) (so y for negative p_z is just -y for positive p_z) that partons 1, 2 interact only if their relative y value $y_1 - y_2$ is of order one or smaller.

(I use 1 GeV for the general energy values at which interactions fall off etc. I suspect in several applications even a smaller value (e.g. $p_\perp$ average) may be correct, although possibly larger in some circumstances - it of course cannot be defined precisely without a quantitative theory.)

We use this assumption to get at the wave function (in terms of parton distributions) for the outgoing final state after interaction. The distribution in y of partons for the in states A and B have ranges of y small to $\ln 2P_A$ and $- \ln 2P_B$ to small respectively. We put them together smearing things (effect of interaction) over a range $\Delta y = 1$. This smearing near $y = 0$ joins the positive and negative y regions (from A and B respectively). Since these regions were the same for both A, B (see A7) this can be done most simply by just extending the common plateau region from one to the other. The exact position of the CM leaves no trace then. We assume this as a general principle.

B2. In a hard hadron-hadron collision there are no special effects that distinguish particles having finite momenta in the exact center of mass. Longitudinal transformations with a velocity v not too close to c leave the distributions of such particles essentially unchanged. (Assumption due to C.N. Yang.)

In our application the particles are partons, the transformation alters the position of the origin of y by $\ell n f = \frac{1}{2}\ \ell n\ \frac{1+v}{1-v}$ (a finite amount) and the assumption says the distribution should look the same. Hence the smearing just has the effect of extending the plateau region of A smoothly back into B.

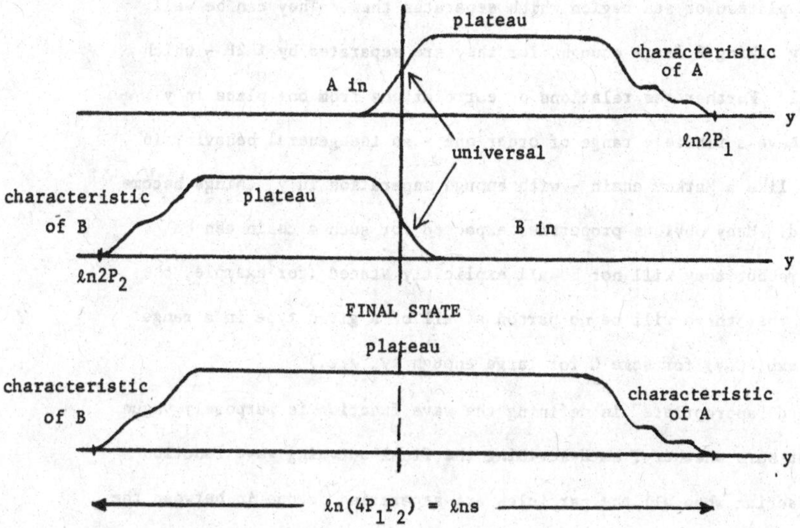

We might ask if the smearing of $\Delta y = 1$ changes the distribution of partons for y near $\ell n2P$ where they were determined entirely by A. (This region is called the A fragmentation region, near $-\ell n2P_B$ the B fragmentation region.) But near A the distribution already satisfies (in the sense that a liquid surface is nearly independent of what goes on deep below if forces extend over finite distances) the wave equation, so we assume it is not changed. Therefore we imagine that the final real hadron particles come from the disintegration of an "original" parton state which has the following properties.

B3. The parton state appropriate to a fast hadron collision of A and B is one whose density matrix looks like that for the A in the A single particle fragmentation region (y near $\ell n 2P_A$, or better if $x = p_L/P$, for $x > 0$), like B alone in the B fragmentation region (y near $-\ell n 2|P_B|$, i.e. for $x < 0$) and like the universal plateau region in regions in between ($|x|$ small and wee).

Thus this is completely described in terms of the wave functions for single particles described in assumptions A1 to A8. We emphasize again that our assumptions are not independent, for example, for B3 to work the plateaus from each particle must be the same as A7 says. We are not trying to develop a logical system of assumptions, but just state a number of mutually consistent (or possibly inconsistent - see quark assumptions later on) ideas.

The picture we are developing in B3 is a wave function like a liquid in the variable y with surfaces A, B at which the distributions are unique but an interior plateau or sea region which separates them. They can be well separated by taking P large enough, for they are separated by $\ell n 2P$ - which is universal. Further the relations or correlations from one place in y to another have a finite y range of order one - so the general behavior in this sea is like a Markov chain - with enough separation in y things become uncorrelated. Many obvious properties expected for such a chain can be expected here but they will not be all explicitly stated (for example, the probability that there will be no parton at all of a given type in a range Δy goes as $\exp(-C\Delta y)$ for some C for large enough Δy, etc.)

The word "appropriate" in defining the wave function is purposely vague for I am not sure whether I am describing the final outgoing wave function after interaction when all the particles are separating or one in between the initial and final. I have not clearly resolved my confusion on this matter - but as I only use the function qualitatively in a manner described in the next lecture I have not had to clarify it.

Technical Footnote on the small momentum region in Assumptions B2 and B3

y near z

On drawing the y plots to describe the parton wave function of assumptions B2 and B3 we assumed the wee region near $y = 0$ where A, B interact, as being completely healed over and just a smooth continuation of a plateau through $y = 0$. This is physically what I want and leads to C2 (next lecture) with uniformity in the y defined there for physical particles and in accordance with invariance under finite velocity transformations which applies to particles. But technically the curve in the y space $y = \ell n\left(\sqrt{p_z^2+1} + p_z\right)$ for partons might show a bump near $y = 0$, a bump which moves when we make a finite Lorentz transformation, which is fine since wave functions need not be relativistic invariants (I thank F. Merritt for pointing this out). But it must be a bump so constructed to have no physical effect as a bump in the final real hadron distribution C2. It is a "theoretical artifact" due to carelessness in finding the right normalization and definition of variables for the wave function. The result cannot really be smooth in the scale of $y = \ell n\left(\sqrt{p_z^2+1} + p_z\right)$ for if so it would not be smooth (near $y = 0$) if I had arbitrarily chosen to use $y' = \ell n\left(\sqrt{p_z^2+1/4} + p_z\right)$ for dy/dy' is not constant.

Lecture 53

Hadron-hadron collisions at extreme energies (continued)

We now go on to describe what the products might look like in hadronic collisions (still leaving out diffraction dissociation). We have, of course, no quantitative way to get from the wave function described in partons to the wave function described in outgoing real hadrons. But we shall simply assume that in y space the relation of parton → hadrons is much like the relation hadrons → partons described in A. Later on we shall have to describe the products expected from states which unlike that in B3, have gaps in them, for example, a state with just two partons at opposite ends of the y scale, separated by 2P. We assume (the "complement" of B3):

C1. The hadrons which would result from the disintegration of a state whose initial wavefunction consists of one parton a(or a few) going to the right at y_a and one b parton going to the left at y_b separated by a large gap $y_a - y_b = \ell n 2P$ consists of hadrons of finite transverse momenta in three regions. There are hadrons going to the right with y near y_a depending on the character of a; there are hadrons going to the left with y near y_b characterized by parton b and those in between distributed in a universal uniform plateau -

like sea which is the "universal sea of hadrons corresponding to a parton gap".

It has been proposed that this "sea corresponding to a parton gap" be itself a gap. This will not be inconsistent with what we will say next (C2), and which we use in hadron collisions. But it does not seem reasonable by itself to me if hadrons make a universal sea as B2 supposes - for I think that means that if there were any distribution of two lumps of hadrons with a gap between, they would make a sea of partons, so if there is no sea of partons there is no separated lumps of hadrons, but there must be a hadron sea. At any rate it is my strong belief that there is in fact such a sea in this case and not a gap in hadron momenta corresponding to the gap in parton momenta.

As A. Cisneros points out the two outgoing lumps carrying opposite hadronic quantum numbers (in the case $e^+e^- \rightarrow$ hadrons) would generate a dipole strong current which would tend to radiate softer hadrons into low x. To prevent this radiation becomes increasingly more difficult as the energy increases, (as for two-body exchange reactions). Any exclusive two-lump probability will fall as a power of energy relative to the total inclusive reaction in which the radiation, generating an intermediate plateau, is permitted.

Now for a wave function as in B3 we can imagine the various partons disintegrating more or less like in C1 but not really independently, those at the ends of the y range determine the hadrons there, and those in the center affecting the hadrons in the center, but in a universal manner independent of y in this region. Thus we again get a hadron distribution like that in C1 with three regions, but the plateau may be a new and different distribution but a sea nevertheless. (Whether the two plateau regions, the one in C1 corresponding to an initial parton gap, and this one for wave function B3, are the same or not is a difficult problem I have not yet been able to decide. We shall call the assumption that they are equal C6, see lecture 55.) This assumption can be made by just repeating the wording of C1 just changing the name of the sea, or it can be put in another totally equivalent way.

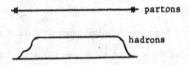

Remark: There is a possible confusion here between the "initial" wave function of C1 and the "appropriate" wave function of B3. C1 in e^+e^- collisions is just after the interaction with the photons - the parton pair is just created. There still remains time for interactions (via terms like $a\,a^*a^*$ in the Hamiltonian) to act before we reach the parton representation of the "final" initial state - i.e. before we reach the "appropriate" wave function in the sense of B3. This interaction converts the initial fast parton into two or more, and these again are broken up etc., in a cascade fashion making profound changes, for example by filling in the gap in the low p_z region and creating some sort of parton plateau in the final outgoing state "appropriate" to the initial state C1.

The reason no such extensive modification is made in going from the initial state in a hadron collision to the final appropriate state B3 is this. The distribution of the fast (non-wee) partons in the initial state already satisfies $H\psi = E\psi$ so little disturbance is worked there by the Hamiltonian. Only in the mutually overlapping wee regions does the further action of H modify things (to smooth out the plateau).

C2. The distribution of final state hadrons at each y resulting from a parton distribution like B3 depends on the nature of the parton distribution within a finite range in y of y.

This is true not because distant partons have no effect, but rather because they have a universal effect. Here (and in C1) y can be more precisely defined for the hadrons are on their mass shell and have a definite mass. We take $y = \ln(E+p_z) = \ln\left(\sqrt{p_z^2+m^2+p_\perp^2}+p_z\right)$ say in GeV (change of scale means just a change in origin of y).

C3. Putting this all together it means that in a hadron collision $A + B \to$ anything in the CM system plotting $x = P_z/P$ for x negative the distribution of products depends only on x and B, for x positive only on x and A as $P \to \infty$; and in the small $\pm x$ region the distribution becomes universal varying like $dx/|x|$ and more generally continuing across the wee region like dp_z/E.

The idea that the right movers depend only on A and the left movers only like B is called limiting (i.e. as $P \to \infty$) fragmentation. It was suggested by C.N. Yang et al, Phys. Rev. 188 2159 (1969) but at the time it was supposed these regions separated and had no communication, but in fact there is a sea between. This sea, however, we suppose is universal and (although it logically could) we assume it carries no information from the right to left region.

We have omitted the diffraction dissociation but it is evident that if it is added in a definite percentage to the inelastic will not change our conclusion. However, the fraction that elastic scattering is of the total cross section does not seem to be universal (for example, for pp at large P $\sigma_{e\ell}/\sigma_{tot} \sim .25$ whereas for $\pi^{\pm}$ p it is closer to .17, see G. Giacomelli in Proceedings Amsterdam Conference on Elementary Particles, North Holland Press (1971)). Thus limiting fragmentation cannot be absolutely exact. It is probably generally nearly correct; perhaps in a future more exact understanding it will be true for parts of the collision characterized by some other parameter (e.g. impact parameter) but when integrated over this parameter it is no longer exact for different cases give various different relative weights to the various values of the parameter. Nevertheless even with this evidence against its perfect universal validity we continue to analyze in a naive and simple way leaving refinements to some future date.

There are a number of additional conclusions made by assuming the Markovian idea and extending our ideas such as about gaps A8 from parton to hadron wave functions. We shall not discuss them in detail for hadron collisions is not our main subject but give some examples. We assume, analogously to A8 that

C4. The probability there are no hadrons in a sufficiently large gap of rapidity Δy goes as $e^{-a\Delta y}$ where a depends on the quantum numbers carried by the gap (right going particle quantum numbers of A minus quantum numbers of all hadrons to the right of the gap). For example, for the exclusive collision A+B → C+D so the outgoing state is pure C to right, D to left, the gap is $\ell n2|P_A|+\ell n2|P_B| = \ell n4|P_AP_B| = \ell ns$ and $e^{-a\Delta y} = (4P_AP_B)^{-a} = s^{-a}$ where a depends on the quantum numbers of A-C. Looked at from a Regge point of view this should go as $s^{-2(1-\alpha)}$ whose α depends on the quantum numbers exchanged in the t channel, which is the same as A-C. Thus a is identified with $2(1-\alpha)$ (or whatever the correct power law of energy fall-off turns out to be) and we make a contact with the theory of exclusive reactions. (The same goes if C is two particles like $\pi + p$ of fixed total mass^2 not necessarily at resonance. I do not know of examples off resonance where this power law has been checked - again we see the universal principle that going to higher energy does not lift a resonance ever higher against "non-resonant" background - the latter can always also be thought of as tails of other resonances.)

Again applying this to the case that C only is near the end of its range, so x_c is nearly 1 but D is anything, even many particles, we see we are generating a gap of $\ln(1-x_c)$ and the amplitude goes as $e^{-a\ln(1-x_c)} = (1-x_c)^a$. The distribution of x_c is then $d(1-x_c)^a = a(1-x_c)^{a-1}dx_c$ with $a = 2-2\alpha$.

This result, though "legally" true as $P \to \infty$, $x_c \to 1$ is in practical cases nearly unobservable. For example suppose C is a proton produced in a p+p collision. Protons also come from diffraction dissociation of a resonance of of mass M_R say going to proton W and pion. This spills protons over a range from $x = (E_p-p_p)/M_R$ to $x = (E_p+p_p)/M_R$ (as $P \to \infty$), where E_p, p_p are the energy and momentum of the proton in the rest frame of the resonance. Although the latter is less than one, there is a very small gap (of range .98 to 1 for $M_R^2 = 2.16$) free of diffraction generated protons it is too small to isolate experimentally. If a gross plot is made for x_c not sufficiently near 1 various numbers of dissociation protons are included and the variation of numbers appears far from $(1-x_c)^{1-2\alpha}dx_c$. This difficulty does not arise for pions when the incident particle is a proton.

Lecture 54

High Energy hadron-hadron collisions (continued)

I should like to make a few comments about our "conclusion" from C2. First since the mean number of particles goes as dx/x in the small region and continues across $x = 0$ as dp/ε it is evident that the total number of particles of a given kind (the multiplicity for that kind) rises logarithmically with $\ln P$ or with $\ln s$. This is also obvious in the area of the y plot where the plateau expands logarithmically with s. But the plateau region is (statistically) neutral, its average for any additive quantum number such as charge, third component of isospin, baryon number, hypercharge, z component of angular momentum etc., must be zero (because if not dx/x would give a ℓns dependent value for one of these fixed conserved quantum numbers). We expect this from the cascade idea of how the plateau is formed. Thus such integrals as $\int_0^1 dx$ [Number of π^+ at x - Number of π^- at x] converge to numbers which are characteristic of the particle A (initially moving to the right) independent of P as $P \to \infty$. Independently corresponding "left numbers" like the same integral for $x = -1$ to 0 can be defined, which should depend on B.

In particular then we can define definite quantum numbers (for the additive quantum numbers) for the right-moving particles, by simply adding the total number for all for which $x > 0$ (i.e. $p_z > 0$) in the CM system. This number will vary from event to event, of course, but we want the statistical expected mean over many events. This "mean right quantum number" will approach a constant as $s \to \infty$. Thus we can talk of the "right mean 3-isospin" or the "right mean strangeness". It is evident that these mean right quantum numbers must, under the ideas of C2, be the same as those of the incoming right-moving particle. The plateau region does not let any quantum numbers slip through it. (If for example we take a symmetric collision A + A then by overall quantum number conservation and symmetry the right quantum numbers (and the left) must be that of A. But by limiting fragmentation replacing the left A by B to make A + B does not change the distribution of right movers hence they still carry the quantum numbers of A.)

Thus interestingly as $P \to \infty$ the right-moving particles in the mean carry the energy, the momentum (minus a constant), the 3 isospin, strangeness, baryon number, z angular momentum, etc., of the incoming right mover.

<u>NOTE:</u> We show that, disregarding quantities of order 1/P, the difference of total energy E and total momentum P of the particles moving to the right is a constant $D = \Sigma(\varepsilon_i - p_{zi})$ (independent of P as $P \to \infty$ and in fact if the plateau is universal, the same constant D for every particle A). For finite x (say positive) the difference from one hadronic particle is $\varepsilon - p = \sqrt{P^2x^2+p_\perp^2+m^2} - Px \approx (p_\perp^2+m^2)/2Px$ (m is the mass of the hadron) which is of order 1/P and therefore negligible. The main contribution comes from x near zero where the distribution of a particular type is $c\,dp/\varepsilon$, hence the contribution to $\varepsilon - p$ of these is

$c\int_{p_z=0}^{\infty}(\varepsilon-p_z)dp_z/\varepsilon$ with $\varepsilon=\sqrt{p_z^2+p_\perp^2+m^2}$. The integral gives $c\sqrt{p_\perp^2+m^2}$, so D is the sum of $c\sqrt{p_\perp^2+m^2}$ over the transverse momenta and the types of hadrons in the plateau. If the plateaus are universal, the constant D is universal and may be easily calculated in terms of already measured quantities.

Thus in a collision of A and B each particle is converted into a train of particles moving in its own direction. The "train A" has the quantum numbers of the particle A and its energy (by the conservation of energy) but has lost a certain momentum D in the interaction, it is held back a bit by the interaction, A and B each lose D to the other. (Such a finite momentum transfer is, of course, consistent and understandable if only wees interact in the collision.)

(For the wave function of a single hadron described in assumptions A2 to A6, the total momentum of the partons is, of course P, the total momentum of the state, but the total energy $\Sigma\varepsilon_i=\Sigma_i\sqrt{p_i^2+m^2+p_\perp^2}$ is not the total energy $E\approx P$ because of interaction energy which compensates the expected finite excess of $\Sigma\varepsilon_i$ above Σp_{zi}.)

As a first step to describe these things formally, we are trying to describe the state

$$|A_{\text{in right}},B_{\text{in left}}>$$

(where "in right" means having very large positive longitudinal momentum P, and "in left" means -P) in terms of outgoing hadron states - an element of the s matrix. Of course the most likely thing is that the two particles do not collide, making simply $|A_{\text{out right}},B_{\text{out left}}>$. We wish to deal with the wave functions if they collide, so we write as usual S = 1 + iT and we are speaking of the T matrix. We shall not normalize it correctly, but just outline our ideas. Formally this wave function can be given in terms of the amplitude to find various outgoing hadrons. If c^* is the (formal) operator to represent the creation of some kind of a hadron (kind, transverse momentum $p_\perp$, and longitudinal momentum p are indices of c^*) we can represent such states by X|VAC> where X is some operator function of the c^*. We have been discussing how X looks. Let M be an operator to create a plateau - say a typical universal plateau for some range of x around 0 - say x = -.2 to +.2

[the exact way the plateau of M cuts off for finite x is arbitrary; its choice affects the definition of G^L, G^R defined later, but the final operator X is not dependent on this]. Next we write X as $G^L G^R M$ where G^R is to modify the sea on the right (for x > 0. It involves creation operators c^* to add particles to (and beyond) the plateau operator M, and annihillation operators c to take particles out (which were put in by our arbitrary choice of how the plateau M is defined) but all these for x > 0 (that is the meaning of the R). Likewise G^L is an operator function of c^*, c only for x < 0. The operators G^L, G^R commute since they contain operators of different particles (some signs must be adjusted for Fermi particles). Thus we write

$$|A_{\text{in }R}, B_{\text{in }L}\rangle = G_A^R G_B^L M|VAC\rangle \tag{54.1}$$

$$= G_A^R G_B^L \,|M\text{-plateau}\rangle$$

where we have written $|M\text{-plateau}\rangle = M|VAC\rangle$. The operator G_Z^R depends only on the particle A, etc. If you want things to look even nicer write the left side in terms of operators too, say d^* which create incoming particles, and then have

$$T d_A^{R^*} d_B^{L^*}|VAC\rangle = G_A^R G_B^L/M\text{-plateau}\rangle$$

Thus the operator $d_A^{R^*}$ is equivalent (in this two-body equation at least) to G_A^R, but in an odd representation in which $d_A^{R^*}$ acts on the vacuum and G_A^R on the M-plateau state.

A research problem which is very important, and virtually unknown theoretically, is (the very rare) collisions at extreme energy in which the particles come out at large relative momenta to the original direction. For example, proton-proton elastic scattering at finite angle, e.g. 90°, where t is the same order as s as $s \to \infty$. What kind of physical view accounts for these collisions, I shall not discuss ideas which have been tried here, for our subject is photons, but shall only comment that nothing is clearly understood and you can start from scratch on your own. (For example will assumption B1 have to be abandoned or quantified?) (You start by looking first roughly at the experimental results to remember qualitative salient features that might need explanation.)

Comments: By assuming that the wee region is the same for each hadron and that only wees interact have we not assumed that all total cross sections σ_{pp} or $\sigma_{\pi p}$ etc. are equal, clearly contrary to fact? I have not thought this out clearly but have always supposed that the part of the wave function which does interact (which is always infinitesimal compared to the part where they go past each other without interacting) could still have some normalization related to the total cross section for that particular collision without being inconsistent with other ideas. In the formal expression above, for example, each G_A could carry a numerical coefficient g_A proper to A. This would make total cross sections proportional to $g_A^2 g_B^2$, or as is said, factorizable. It may be that the previous assumptions do not imply that the total cross sections are necessarily equal, but rather perhaps that they are factorizable. It would imply for example, that $\sigma_{\pi p} = \sqrt{\sigma_{pp}}\ \sqrt{\sigma_{\pi\pi}}$ etc. We do not have any evidence on whether this is true.

In these studies we have made no remarks which permit us to understand transverse momentum behavior (except to say that transverse momenta in hadronic collisions are limited, a result taken directly from experiment). Obviously lots of interesting theoretical questions remain, such as what function is the transverse momentum distribution, how does it differ for various values of x, or for π's and K's? How should exclusive cross sections vary with t, etc.? This entire realm of phenomena has been left out of our analysis, an excellent future opportunity for advance lies here.

Final Hadronic States in Deep Inelastic Scattering

Lecture 55

Interaction of partons with the electromagnetic field

We assume that in the original field Hamiltonian describing hadrons in terms of partons there are terms giving the coupling of partons with the vector potential of the quantum electromagnetic field. We shall assume in the spirit of minimum electromagnetic coupling that they couple in the simplest way expected from the propagation operators via gauge invariance. That is we assume:

D1. The coupling of partons to the electromagnetic field is the ideal minimum coupling operator. That is also the coupling that would be valid if they were ideal free particles.

This coupling is not unique if the partons are spin 1 or higher, but for the present this will not concern us for we shall suppose partons are either spin 0 or spin 1/2. Although we are in danger of not having the most general case we shall nevertheless explicitly next take the working hypothesis (suggested, of course, as we have discussed by experiment on νW_2 and W_1 and not a priori by theory) that

D2. All the charged partons are of spin 1/2 and hence couple through the current operator $e_\alpha \gamma_\mu$ where e_α is the parton charge (α is an index for the kind of parton).

We have seen how the assumptions A1-A8 plus these two D1 and D2 lead to the scaling expectations for the deep inelastic scattering (lecture 27) and there is no reason to repeat all that here again. However, here we shall discuss what we can say about the products in photon collisions, in particular we begin with the deep inelastic ep scattering region $q^2 = -2M\nu x$, $M\nu = P\cdot q$ (P is the momentum of the proton, q that of the virtual photon) so ν = virtual photon energy in laboratory (proton at rest) system. Let us use the coordinate system with the virtual photon purely spacelike $q_\mu = (0, -2Px, 0, 0)$ $P_\mu = (P, P, 0, 0)$ $q^2 = -4P^2x^2$, $2\,M\nu = 4P^2x$. Then as a result of our awsumptions the parton wave function before and after the collisions looks like:

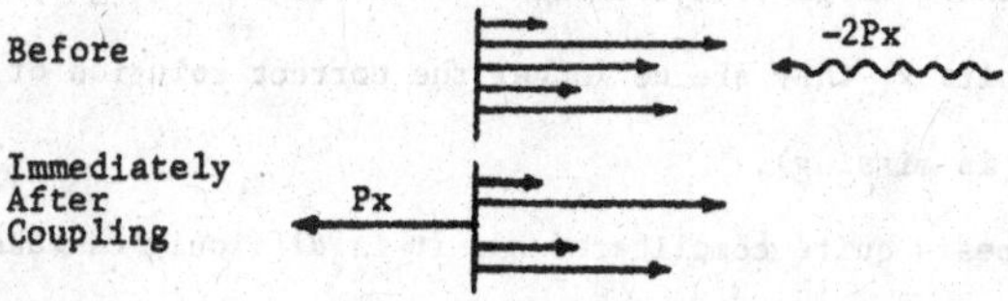

That is, one parton (say type α) moving to the left, the remaining partons moving to right just as in original proton, less the individual parton α of momentum x. The relative probability of this picture is $e_\alpha^2 n_\alpha(x)$ where $n_\alpha(x)$ is the number of partons of type α with $p_z/P = x$ in the original proton state. Then: the hadron products in deep inelastic region are those expected from the parton distribution just described. The total cross section in this scale is proportional to $\sum_\alpha e_\alpha^2 n_\alpha(x)$ so the density matrix per collision is a superposition of cases of different types α of partons with weights, W_α depending on the character of the collision through x; $w_\alpha(x) = e_\alpha^2 n_\alpha(x)/\sum_\beta e_\beta^2 n_\beta(x)$; the sum of the weights being 1.

One obvious consequence of this and our other assumptions is that in this system the transverse momenta will be limited and (for fixed x as we vary P, or if you like ν).

The longitudinal momenta will scale as P, i.e. if they are stated in units of P as say ηP the distributions will be independent of P as $P \to \infty$ (depend only on η).

We expect also near η equal zero to find a $d\eta/\eta$ behavior. For positive η (night movers) we expect it behaves like the universal M-plateau characteristic of a wave function like B3, which we know from hadron collisions. For negative η we are in the "plateau region of initial parton gap" (defined in connection with C1). We have not assumed these two plateaus are the same so the coefficient of $d\eta/\eta$ need not be the same. If they are not we shall have trouble defining what happens in the transition region - it cannot go simply as dp_z/ϵ for that leads to the same coefficient for plus and minus η. We see however that this question is to some extent an artifact of our particular choice of coordinate system. Note that the state described here as "immediately after coupling" is an initial parton state (in the sense discussed in the remark following the discussion on C1) - there still must be interactions from the Hamiltonian before it becomes the "appropriate" outgoing wave function. This will produce cascading of the left-moving parton into the gap smearing the wee region into negative η and making large readjustments for the right-moving system also (because, for finite x, they are no longer the correct solution of $H\psi = E\psi$ since one parton is missing).

This all appears quite complicated and it is difficult to make firm predictions. However, we might continue to assume interactions are each limited in range on a rapidity plot - although there are many of them possible filling in gaps, etc. But we shall try to adhere to the principle at least that the parton moving to the left determines the final hadrons to the left and likewise for the right. We put this idea formally into the following assumption, a generalization of C1, C2 (we write it independently, for it may not be true while the special case C1 or C2 may be).

C5. In the center of mass system (or one moving longitudinally at any velocity not near c) an initial state consisting of any distribution of partons to the right and other partons to the left yields in the final state a distribution of hadrons such that those hadrons moving to the right are entirely characterized by the initial partons moving to the right, and does not depend on those initial partons moving to the left. (Likewise, exchanging left and right.)

Assumption C5, if it were right and this continuity in dP_z/ϵ, would seem to suggest that both plateaus fit together. I am not sure of myself here but shall put it down as an explicit assumption which would, if it is true, remove all our difficulties - the $d\eta/\eta$ region is always universal, that of hadron-hadron collisions.

C6. The distribution of hadrons in the plateau region is the same for every initial parton distribution. This assumption is at present, very weakly based and may easily be wrong - it is an interesting conjecture.

We now make a more detailed discussion of our expectations for the left-moving particles. (I have profited greatly from conversations with A. Cisneros on these matters.) For these particles a variable more convenient than η (which goes down to $-x$) is $z = -\eta/x = -p_z/Px = P\cdot p/P\cdot q$ the fraction that the left-moving particle's momentum is of the total left-moving momentum. Since this is $z = P\cdot p/P\cdot q$ it is the energy of the particle in terms of the energy ν of the photon in the laboratory system. It is the proper variable for seeing how the virtual photon fragments. Of course as $\nu \to \infty$, x fixed, the distributions in z scale as we have said.

If we could be sure that only a parton of type α came out (which by the way, can be much more nearly done for neutrino scattering - in the quark model, neutrino scattering can lead to unique quarks to the left) the distribution to the left would be unique - say $D_\alpha(z)$. The number of π's with a given z is a function of z only $D_\alpha^\pi(z)$, the probability of π, K at z_1, z_2 depends on z_1, z_2 as $D_\alpha^{\pi K}(z_1, z_2)$ - and these functions do in no way depend on x. They do not depend on x because the hadrons to the left depend only on the parton to the left (α) and the adjacent wees from the hadron (if at all) - and these latter are universal and unaffected by the removal of the parton at x from the proton. The latter does not affect left-moving hadrons for its relative momentum to left-moving hadrons is not finite but grows as $P \to \infty$.

The actual distribution seen at a given x will depend on x because the relative probabilities of producing different kinds of partons α will depend on x. The actual distributions D(x,z) will be the weighted average for each parton

$$D(x,z) = \sum w_\alpha(x)\, D_\alpha(z) \tag{55.1}$$

where the weights $w_\alpha(x)$ proportional here to $e_\alpha^{\;2} n_\alpha(x)$ are defined by

$$w_\alpha(x) = e_\alpha^{\;2} n_\alpha(x) \Big/ \sum_\beta e_\beta^{\;2} n_\beta(x)$$

These functions $D_\alpha(z)$, or equivalently their creation operator D_α^R on M-plateau (if C6) isolates something characteristic of partons and, if our assumptions are all correct, therefore very fundamental indeed. We shall discuss later a specific parton model (quarks) as well as some practical questions about the possible extraction of the isolated $D_\alpha(z)$ from experiment, as well as the possibilities of finding the $w_\alpha(x)$ by special guesses about how $D_\alpha(z)$ may behave. To me the possibility of special functions characteristic of each kind of parton is a very interesting possibility, and one that could be an entrance to a path into the heart of the mechanisms of strong interactions.

These same functions $D_\alpha(z)$ will appear in certain other experiments - for example, of course, in deep neutrino proton → μ + products experiments. The analysis is nearly the same as here except that the fundamental coupling may be different so although $n_\alpha(x)$ are the same the weights $w_\alpha(x)$ come out differently.

Again in the e^+e^- collision the assumption D2 says our initial state is just a pair, parton α and antiparton $\bar{\alpha}$ with weight $e_\alpha^{\,2}$. Thus, assuming C6 the final state would be $\sum_\alpha e_\alpha^{\,2}(D^R_\alpha D^L_{\bar{\alpha}} + D^R_{\bar{\alpha}} D^L_\alpha)|$M-plateau> again producing hadrons in any one direction characterized by the distribution

$$\sum_\alpha e_\alpha^{\,2} D_\alpha(z)$$

where we sum an α over partons and antipartons.

If, for example, in some experiment we could be sure that a certain parton α came out to the left say, then as we have seen we would expect that the total "left-moving quantum number" (the sum, for some additive quantum number of that number for all hadrons moving to the left averaged over all events) would be that of the parton α. Thus in principle we could define or determine in terms of experiment the quantum numbers of the partons. If the state is not pure we shall have to know something of the weights $w_\alpha(x)$ to make this useful - but there are so many different kinds of experiments possible that in principle the $w_\alpha(x)$ can be determined as well as the overall quantum numbers of the partons.

The particles to the right under deep inelastic e-p scattering come from fragmentation of a proton with one α parton of momentum fraction x removed, say $E_{(p-\alpha,x)}(z)$. They are evidently not very fundamental. But it is clear that the same kind of final state results (on both sides, left and right) in Drell's experiment $p + p \rightarrow \mu^+ + \mu^-$ + any hadrons so the products in this experiment can be entirely expressed in terms of these $E_{(p-\alpha,x)}$, and hence (supposing $n_\alpha(x)$ has been worked out) in terms of the products for deep ep scattering. We leave it for you to write the explicit relations and to suggest practical experiments to test your ideas.

NOTE: Our final hadron state is according to our assumptions

$$\sum_\alpha w_\alpha(x)\, D_\alpha^L\, E^R_{(p-\alpha,x)}|M \text{ plateau}\rangle$$

where D_α^L is the operator for a left parton α, and $E^R_{(p-\alpha,x)}$ for the right-moving fragments. But this can be considered as a mnemonic only for the expression is probably impossible for one operator M always. Because nothing would seem to prevent us from writing $D_\alpha^L\, D_\beta^R|M \text{ plateau}\rangle$ which would have total quantum numbers of two quarks (α + β) which is impossible to write in terms of the hadron operators having only integral quantum numbers. (I am indebted to J. Mandula for pointing this out.) A valid mathematical representation for these ideas is an excellent problem.

The reader should be warned that a number of these scaling predictions for special products of reactions may only hold at much higher energies than that at which scaling for the total cross section (νW_2 and W_1) sets in. This warning results from theoretical experience with a number of examples of analogous theorems in non-relativistic quantum mechanis where the sum works well before the individual terms do. This is because if certain interactions are disregarded in working the total probability by assuming certain states only are "entered" subsequent interactions may not change the total probability the state was "entered" but may redistribute that probability over different final states than were expected.

Special case of small x

In the special case of small x the predictions are especially simple. First consider the right side (original proton). Here we have a parton distribution just like that of a proton with only a very low x parton removed and the wees disturbed (by interaction with the plateau developing from the left). Thus all the partons ot any substantial x are exactly like that of a proton - and we can expect the same distribution of hadrons to come out (at least for z >> x) as do come out for a hadron collision of a proton, say

F_p^R or $F_p^R(z)$. Hence for x small $E^R_{(p-\alpha,x)} \approx F_p^R$.

Next for small x all $n_\alpha(x)dx$ go as $C_\alpha\, dx/x$ where C_α is a constant, so that $w_\alpha(x) = C_\alpha/\sum_\beta C_\beta = \gamma_\alpha$ approaches a constant γ_α independent of x for small x. Next call D_Γ^L the mixture $D_\Gamma^L = \Sigma\gamma_\alpha D_\alpha^L$ of the distributions for each parton, each weighted with weight γ_α. Our hadron distribution becomes thus, for small x, nearly

$$D_\Gamma^L\, F_p^R|\text{M-plateau}\rangle$$

That is, for small x the proton fragments into a form independent of x and the same as it does for a hadron collision. And the virtual photon also fragments in a universal way independent of x. Since we have assumed the low x region the same for all hadrons, the C_α and γ_α, and hence D_Γ do not depend on the particle struck by the photon (normalized to the total cross-section for collision, of course). A small x photon and a hadron behave just like the collision of two hadrons, each fragments in its own characteristic way. That of the photon is independent of x.

Region of finite q^2, $\nu \to \infty$

For finite q, negative q^2 we can still use our system of coordinates in which q has only a space component Q.

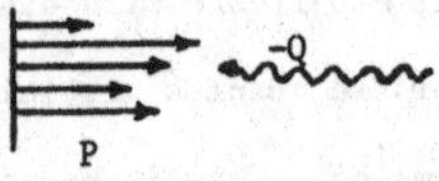

except if Q = 0. It is clear here however that only x near zero can be affected by the photon Q; that is only the wees are effected. They are, however, affected in a very complicated way for interaction is important in the wee region. We cannot therefore predict what will happen there, but we can note (a) that it is the same for every hadron A, A + $\gamma \to$ products for we have the same wees for every hadron according to A6, and (b) the fragmentation of the finite x in the above system is characteristic of partons of system A only, for only the wees are effected by the photon.

In consequence of (a) the products on the left which can be described in terms of $z = \frac{P\cdot p}{P\cdot q}$ where P is the proton four-momentum, p is that of a product and q that of the photon, for finite z as $\nu \to \infty$, is some kind of a distribution $D_{\gamma,q^2}(z)$. The distribution clearly depends on q^2, the virtual mass of the photon, because the complicated interactions of the wees depend on this momentum. In the other direction (the variable $(q\cdot p/q\cdot P)$ the proton fragments in the same way as it does for hadron collisions. These considerations hold for $q^2 = 0$ also, of course, but our coordinate system is inconvenient for such a case.

We could also use the center of mass system for any finite q^2

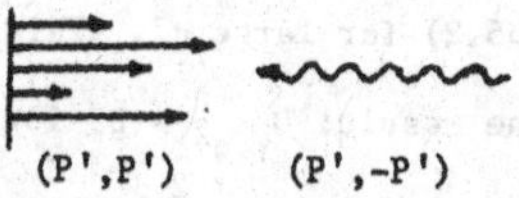

Conservation of energy and momentum means the virtual photon ($P' \to \infty, q^2$ finite) interacts only with the wee partons of the target proton (or hadron A). This interaction is complicated but produces the same distribution for any hadron for given q^2. The hadron behaves as it always does where its wees are disturbed whether by another hadron or by a photon. Formally our final hadron state is

$$D^L_{\gamma,q^2} F^R_A |\text{M-plateau}\rangle \tag{55.2}$$

Thus as far as high energy inelastic collisions are concerned the (virtual or real) photon acts just like a hadron inasmuch as it appears to have its own (q^2 dependent, or $q^2=0$) fragmentation products, in its direction, the hadron fragmenting also in its characteristic way.

This of course makes a nice union with the idea of vector meson dominance, that a free photon ($q^2=0$) has a certain reasonable probability to be a virtual vector meson and as such would behave in hadron collisions like a hadron. We note now we shall <u>not</u> have to determine with what probability it looks like a hadron and how this varies with q^2, for in any event it, as a whole, should act just like a hadron does in $\nu \to \infty$ collisions.

In the center of mass picture (and also in the spacelike q figure) there are terms of coupling in which the photon first divides into partons on the way in, for example one fast one slow, and these slow partons interact or annihilate with the wee partons of the hadron. Thus the picture that the incoming photon looks with some amplitude like partons itself is reinstated. As q^2 rises (and certainly where $x = -q^2/2M\nu$ is finite) the contribution of such diagrams falls away and only the direct coupling term of photon scattering a parton of the hadron remain important.

Continuity of large q^2 and small x region

Finally we shall match our finite q^2 region to our small x region. As we have done before we shall suppose when ν is very large and $-q^2$ large but $-q^2/2M\nu$ small the limit may be taken in either order - i.e., we can get the result either from our finite q^2, $\nu \to \infty$ formula or from x finite, but small, $\nu \to \infty$ formula.

Thus (55.1) must agree with (55.2) for large q^2. This is easily done, the results agree if only we add the result: $D_{\gamma,q^2} = D_\Gamma$ for large q^2. That is: The fragmentation products of a photon of large $-q^2$ become independent of $-q^2$ as $-q^2$ rises. (We are in all cases normalizing to the total cross section which is varying, as $1/q^2$, of course.)

Partons as Quarks

Lecture 56

Partons as Quarks

We could now go on to discuss various models of what quantum numbers partons carry, but we shall limit ourselves to one example, the one that is most interesting. The student should try other examples, such as the Sakata model, to see whether we can eliminate them by experiments now done, or proposed.

We shall suppose the charged partons come in six varieties, three plus their antiparticles. The three called u,d,s carry the quantum numbers of the three quark states (of the low energy quark model). This we summarize by:

E1. The charged partons are quarks. Most of our previous assumptions were guided, or we thought they were guided by field theory or considerations based on high energy experiments. This, of course, is not, it is an inspired guess. But it is also contrary to what can be true if the field theory is too ordinary. For in such a theory there would be a base state of quark number one (and non-integral charge in a localized wave packet) and, in view of the conservation of quark number, some eigenstate of the system of quark number one is expected. In other words we expect to see real particles with quark quantum numbers. They have not been seen. It is possible to imagine that they have very large mass - but this makes it very hard to take all the previous assumptions about all parton interactions limited to the GeV region, etc. There may be some way to reconcile all this - it is one of the most intriguing of theoretical problems. In order to emphasize it I will make another unnecessary assumption that I will not use but I add to remind you of the problem.

E2. Physical quarks do not exist. If you prefer to replace it by "physical quarks have high mass", go ahead - you still have theoretical work to do reconciling it with E1 and the rest of our assumptions. Of course E1 may be wrong - one of the most important experimental jobs of the future is to find out whether E1 is indeed correct or impossible and so we should work out as many testable consequences of it as possible. So far we have, as discussed in lecture 32, come to the conclusion from experiment that in any case:

E3. Neutral partons also exist. What they may be like we do not now know, except perhaps they may not be vector (because the ω,ρ degeneracy is not lifted).

Although the problem of reconciling E1, E2 and field theory might be very difficult, it appears (at least at first sight) to be not at all difficult to reconcile E1, E2 and the other assumptions we have explicitly made about parton and hadron distributions!

A careful review of our assumptions shows this. There is possible doubt, as already mentioned, in B1 (interaction only between partons of small relative y) but this is only used to make the later assumptions more plausible and maybe replaced by these latter assumptions.

A place where there is an especially interesting, but not inconsistent, conclusion is in connection with "right-moving quantum numbers" (discussed in lecture 54) for the parton function D_α^R and which should now be non-integral. These numbers are defined statistically as the average over all events - although each event must give integral values the average, of course, need not be. For example if we know one quark (and no antiquark) is sent to the right, the mean number of baryon less antibaryons found on the right should (at extremely high energy, at least) be + 1/3.

The arguments leading to this conclusion are not obviated by the fact that the partons do not have integral quantum numbers. Imagine for a very large ℓnP that the quarks are distributed in the (final state) wave function in a long plateau: (the left end of the plateau is generated by a long chain of cascades via terms like $a^* a^* a$ from the initial single quark)

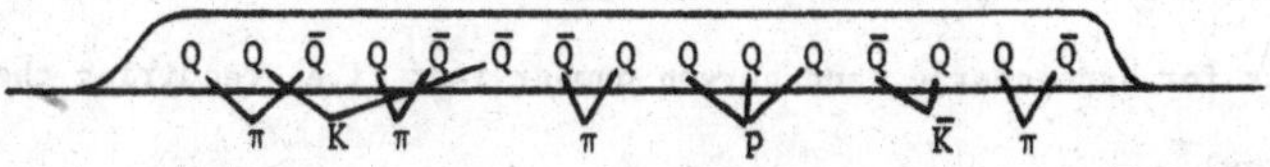

Then in turning into final hadrons various bundles of quarks go together to make legitimate hadron quantum numbers. In doing so they take combinations of quarks over a finite range of y as illustrated above. (The overall total triality must be zero, of course, for the initial state has zero triality; in our illustration the initial state has baryon number one.) We assume that there is some non-zero probability per dy to pick of 3 quarks (or 3 antiquarks) to make a baryon (or antibaryon). Likewise there should be a finite probability to pick up strange quarks. It is then seen that as a statistical matter with a sufficiently long plateau in y (sufficiently long Markovian chain) the quark number (and strangeness) becomes randomized and the central region of the plateau is neutral on the average in these variables. This means that the right and left mean quantum numbers of the final hadrons approach a constant that depends as we have supposed on the initial right or left quark character. It doesn't make any difference exactly where you cut the plateau in deciding which is right or which left, as long as you cut somewhere near the middle.

This result is so interesting and its experimental verification would represent such a direct measure of the supposed non-integral quantum numbers of the parton quarks that we should say some words about its possible verification. First in electron production in general we do not have a single type of quark thrown to the left so the beauty of the result would be confused by having to first know the $w_\alpha(x)$ by one method or another (see methods below). On the other hand for x near 1 we have been led by experiment (ratio of νW_{2n} to νW_{2p}, lecture 31) to suppose that only u quarks survive - hence if that is true, near x = 1 our left-moving quark may be a pure u quark. Another way to insure pure quarks is by neutrino scattering which we discuss below.

Secondly, although it may be easy to form pions in the plateau region it may be harder to form K's and still more difficult to form baryons (because of their masses?). If this is the case we should need a very long plateau indeed to get equilibrium in baryon number, although hyperon charge might be easier and isospin easiest of all. In the energy ranges available to experiment therefore I would expect the quantum number rules to work best for isospin, next for hypercharge, and baryon number last (i.e. requiring the largest energy).

The easiest would be isospin. For example we would expect that the left-moving z isospin: $\Sigma I_{zi}N_i$, the sum of each product hadron moving to the left (for example in the system with q_μ pure spacelike, or maybe the center of mass), will per collision, for a given x be given by

$$\sum I_{zi}N_i = \left[\frac{1}{2}\left(\frac{4}{9}\,(u(x) - \bar{u}(x))\right) - \frac{1}{2}\left(\frac{1}{9}\,(d(x) - \bar{d}(x))\right)\right]/f^{ep}(x). \qquad (56.1)$$

that is, it will rise from zero for small x (where $u(x) = \bar{u}(x)$ etc.), never rise above + 1/2 (or fall below - 1/2) and approach + 1/2 for x near 1 (where, we think u(x) dominates).

This relation can serve as can a number of others of this type in one of two ways. In one case we might suppose u(x), $\bar{u}(x)$ etc., already known from some other method - such as neutrino scattering explained in lecture 33 - or analysis of $p + p \rightarrow \mu^+ \mu^-$ + anything, or by perhaps one of the equations developed further on. In that case (56.1) is a quantitative prediction to test the consistency of all the ideas that partons are quarks. Alternatively it can itself be used to give further information on the six separate functions u(x), $\bar{u}(x)$ etc., so that they may be separately determined. These could then be compared to the results of other methods of getting them, but do not serve directly as a check of the model.

However, even without complete separation of all the functions a relation such as (55.1) may be used to check the quark model because of the existence of sum rules (Equation 31.2) such as $\int_0^1 (u(x) - \bar{u}(x))dx = 2$, $\int_0^1 (d(x) - \bar{d}(x))dx = 1$ so the integral over x of the numerator of (56.1) should be 7/18.

I do not know at this time which kind of experimental information will become available first and so my general discussion suffers from a confusion in analyzing these theoretical expectations as to which comes first the horse or cart. There are a very large number of such relations whose organization is thereby made difficult. I will therefore merely indicate some of the general relations expected from our theory and leave the choice of the best way to use them or combine them to compare to experiment up to you.

Lecture 57

Partons as Quarks (continued)

There is another way to insure that one gets a pure quark of one type recoiling in the final state, and that is with deep inelastic neutrino or antineutrino scattering, as we discussed in lecture 33. We assume of course, the usual $GJ_\mu^* J_\mu$ weak coupling with the hadron part of the weak current given in terms of quarks as Cabibbo suggested. That is, we explicitly assume:

E4. The weak interaction is via the current $\bar{Q}'\gamma_\mu(1+\gamma_5)Q$ where Q is the Dirac operator for a u quark and Q' that for a "Cabibbo quark" - a quark which is d with amplitude $\cos\theta_C$ and s with amplitude $\sin\theta_C$ ($\sin\theta_C \approx .24$).

We shall assume for our discussion that the current interaction is point-like, but that is a matter for experiment to decide. This very interesting question is, however, beside the scope of our discussion; after it is determined the same kinds of questions and remarks will apply to products generated by the aforementioned current operator. In all cases we can analyze things as if they were the effect of a virtual W meson field W_μ of momentum q_μ (generated by the lepton) coupled with the current in E4. Just as before (Equation 33.2) the total square matrix element of J_μ can be expressed in terms of W_2, W_1, W_3 so the products, summed over spins and angles, at least, can also be so analyzed. That is the total W_1, W_2, W_3 can each be split into partial W_1, W_2, W_3 for special types of products. We discuss only the scaling region. We have seen νW_2 is related to W_1 and the relation should hold for products also. But more interesting we noticed $f_1^{\bar{\nu}p}-f_3^{\bar{\nu}p}$ (just the scaled function $2MW_1$-$2MW_3$ for antineutrino scattering on protons) was just purely u(x). Hence the same holds for the products, ie., the probability of a product in W_1 (determined as the appropriate coefficient in the variation of cross section with laboratory neutrino angle keeping q, ν fixed) minus that in W_3 for $\bar{\nu}p$ scattering are purely the products for a u quark being knocked backwards (forwards is a proton less a d quark with probability $\cos^2\theta_C$ - the latter can in first approximation be neglected as $\sin^2\theta_C$ is only .06). Thus by studying the products to the left in this combination we are studying the fragmentation products, $D_u(z)$, expected of one single quark, a u quark in fact.

By choosing other combinations we can select recoil quarks of different types. For example, $f_1^{\nu p}+f_3^{\nu p}$ for neutrinos on protons gives pure $\bar{u}$ quarks and should give various products with probability $D_{\bar{u}}(z)\bar{u}(x)$; a fixed distribution independent of x, the total cross section being $\bar{u}(x)$. Again $f_1^{\nu p}-f_3^{\nu p}$ for neutrinos on protons gives <u>nearly</u> a pure d quark ($\cos^2\theta_C = .94$ d quark, $\sin^2\theta_C = .06$ s quark).

From these distributions the total quantum numbers of the quarks can be determined. In this way alone certain models can be eliminated, but others can not. For example we cannot distinguish the quark model from the three-triplet model where there are 9 partons (and 9 antipartons) in sets of three. A, B, and C each set having three states like quarks with various integer quantum numbers. Thus what would come out in the experiment on f_1-f_3 for νp scattering, where we expect a pure u quark in the quark model, would, in the three-triplet model, be a + 1/2 isospin parton but either of type A or B or C with equal probability so the mean charge on other numbers can be 1/3 integral and just equal to the u value, for that is how the integer charges for A, B, C were chosen. (Other experiments, such as $e^+e^- \to$ anything or correlations of left and right side disintegrations might distinguish these models.)

<u>Product predictions</u>

There are a large number of predictions implicit in the relation that the distribution of a given final hadron in the left (photon) direction (for convenience not normalized) is in general given for deep inelastic ep scattering by: (Equation 55.1)

$$\begin{aligned} D(x,z) &= \sum e_\alpha^2 n_\alpha(x)\, D_\alpha(z) = \\ &= \frac{4}{9}\, u(x)D_u(z) + \frac{4}{9}\,\bar{u}(x)D_{\bar{u}}(z) + \frac{1}{9}\, d(x)D_d(z) + \frac{1}{9}\,\bar{d}(x)D_{\bar{d}}(z) + \\ &+ \frac{1}{9}\, s(x)D_s(z) + \frac{1}{9}\,\bar{s}(x)D_{\bar{s}}(z) \end{aligned} \tag{57.1}$$

where $D_u(z)$ etc., are the distributions of the product in question for pure up quarks, etc. There are six functions in general so it is difficult to analyze unless u(x) etc., were all already available. However, by taking certain combinations of measurements fewer functions are involved. We illustrate this with an example.

Suppose we ask to produce a π^+, call the distribution $D^{\pi^+}(x,z)$ and $D_u^{\pi^+}(z)$ etc. By isospin reflection the probability a u yields a π^+ is the same as that a d produces a π^-; and by charge conjugation again the same as that a $\bar{u}$ produces a π^-. In this way we see for π production there are really only three independent functions

$$D_u^{\pi^+} = D_{\bar{d}}^{\pi^+} = D_{\bar{u}}^{\pi^-} = D_d^{\pi^-}$$

$$D_d^{\pi^+} = D_{\bar{u}}^{\pi^+} = D_{\bar{d}}^{\pi^-} = D_u^{\pi^-}$$

$$D_s^{\pi^+} = D_{\bar{s}}^{\pi^+} = D_s^{\pi^-} = D_{\bar{s}}^{\pi^-} \tag{57.2}$$

In fact if we measure the number of π^+ minus the number of π^- at a given z it (Equation 57.1) all reduces to one function:

$$D^{\pi^+}(x,z) - D^{\pi^-}(x,z) = A(z)\left[\frac{4}{9}\,(u(x)-\bar{u}(x)) - \frac{1}{9}\,(d(x)-\bar{d}(x))\right] \tag{57.3}$$

(for virtual γ on proton)

where $A(z) = D_u^{\pi^+}(x) - D_u^{\pi^-}(z)$. Thus we expect the distribution (probability as a function of z) to be the same for all x. As we vary x we can determine $\frac{4}{9}(u(x)-\bar{u}(x)) - \frac{1}{9}(d(x)-\bar{d}(x))$, within a constant. This is just as in 56.1, but we do not have to measure over all z to integrate, and measure other particles as well. A mere measurement of π^+ and π^- at some convenient z would be enough. The absolute coefficient can be determined in two ways, either from the sum rules (Equations 31.2), or by the hypothesis that as $x \to 1$ only u(x) survives as $\frac{4}{9}\,u(x) \to f^{ep}(x)$ a known function as $x \to 1$. Additional information would come from the same experiment on the neutron, of course, (we get $\frac{4}{9}\,(d-\bar{d}) + \frac{1}{9}\,(u-\bar{u})$).

The sum of the number of π^+ and π^- does not give us much that is new about the distributions, but we can roughly predict its x dependence

$$D^{\pi+}(x,z) + D^{\pi-}(x,z) = (D_u^{\pi+}(z) + D_u^{\pi-}(z))\left\{\frac{4}{9}\,(u(x) + \bar{u}(x)) +\right.$$

$$+ \frac{1}{9}\,(d(x) + \bar{d}(x)) + \frac{1}{9} - \frac{D_s^{\pi}}{D_u^{\pi^+} + D_u^{\pi^-}}\,(s(x) + \bar{s}(x))\Big\} \qquad (57.4)$$

The expression in curly brackets is the same as $f^{ep}(x)$ except for the coefficient of the last term (which should be simply $\frac{1}{9}$). However, that term is probably small (for not only should $s,\bar{s}$ be less than say $u,\bar{u}$ in a proton but also u is enhanced by $\frac{4}{9}$ relative to $\frac{1}{9}$) so the distribution of π^+ plus π^- is probably nearly independent of x, and if normalized to $f^{ep}(x)$ depends on z only.

Arturo Cisneros has suggested an hypothesis which we explain in more detail below, which amounts to assuming that near z = 1 the functions $D_\alpha(z)$ fall off with various powers of (1-z) and in particular that as $z \to 1$ the function $D_u^{\pi^+}$ is much larger than $D_{\bar{u}}^{\pi^+}$ or $D_s^{\pi^+}$. This makes the coefficients of (57.3), (57.4) equal as $z \to 1$. Thus it means that in this region the probability of finding a π^+ as we vary x is a direct measure of $u(x) + \bar{d}(x)$, and the probability of finding a π^- measures $\bar{u}(x) + d(x)$ in the same scale. This is still another suggestion of how the individual determination of the functions $n_\alpha(x)$ may be facilitated. In fact, this hypothesis, if true, permits a determination of the six functions $u(x)$, $\bar{u}(x)$ etc. (up to an overall numerical constant), by measuring the distribution functions for charged mesons only near z = 1, for both proton and neutron. If only the proton is used as a target the $n_\alpha(x)$ cannot be determined without making measurements of neutral mesons which is difficult experimentally.

We can do similar things for the production of other particles, for example K-mesons. Here there are six independent functions $D_\alpha^{K^+}(z)$, the others are obtained by isospin reflection or charge conjugation. (For example, $D_u^{K^+} = D_{\bar{u}}^{K^-} = D_d^{K^0} = D_{\bar{d}}^{\bar{K}^0}$. The student can verify that for ν on protons if we measure particles at a given z to the left, if $N^+(z)$ is the number of K^+ etc., we find the following results: The isotopic spin difference depends on one function, with the same x dependent coefficient as before:

$$N^+ - N^0 + N^{\bar{o}} - N^- = \left(D_u^+ - D_u^0 + D_u^{\bar{o}} - D_u^-\right)(z)\,\left(\frac{4}{9}\,(u(x) - \bar{u}(x)) - \frac{1}{9}\,(d(x) - \bar{d}(x))\right),$$

another combination (which does not require K^0 and $\bar{K}^0$ to be distinguished) also factorizes into one function of x, one of z:

$$N^+ - N^0 - \bar{N}^0 + N^- = \left(D^+ + D^0 + \bar{D}^0 + D^-\right)(z)\left(\frac{4}{9}\,(u(x) + \bar{u}(x)) - \frac{1}{9}\,(d(x) + \bar{d}(x))\right).$$

The sum of all four depends strictly on two functions of z:

$$N^+ + N^0 + \bar{N}^0 + N^- = \left(D_u^+ + D_u^0 + \bar{D}_u^0 + D_u^-\right)\left\{\frac{4}{9}\,(u+\bar{u}) + \frac{1}{9}\,(d+\bar{d}) + \left(\frac{D_s^+ + D_{\bar{s}}^+}{D_u^+ + D_u^0 + \bar{D}_u^0 + D_u^-}\right)\frac{1}{9}\,(s+\bar{s})\right\}$$

but the last factor in curly brackets is likely to be close to $f^{ep}(x)$ for any z. Finally the fourth relation, involving two functions is

$$N^+ + N^0 - \bar{N}^0 - N^- = \left(D_u^+ + D_u^0 - \bar{D}_u^0 - D_u^-\right)\left[\frac{4}{9}\,(u-\bar{u}) + \frac{1}{9}\,(d-\bar{d})\right] + \left(D_s^+ - D_{\bar{s}}^+\right)\frac{1}{9}\,(s(x) - \bar{s}(x))$$

(a measure of hypercharge, but unequally sensitive to strange and non-strange quarks since we do not assume SU_3 invariance).

Cisneros' assumption E6 (below) here means that, as $z \to 1$ only D_u^+ and D_s^+ survive, call them α, β respectively. Then a measure of K mesons to the left as $z \to 1$ is a direct measure of various combinations of the u,d functions.

Number of $K^+ = \alpha u + \beta\bar{s}$, Number of $K^0 = \alpha d + \beta\bar{s}$

Number of $K^- = \alpha\bar{u} + \beta s$, Number of $\bar{K}^0 = \alpha\bar{d} + \beta s$

In this, as in all cases, data on νn gives additional information, change $u(x) \leftrightarrow d(x)$ and $\bar{u}(x) \leftrightarrow \bar{d}(x)$ in the formulas.

We have not discussed the right distributions, but there are relations here too for various experiments. We mention only one as an example. Llewellyn Smith's sum rule (Equation 33.6) neglecting $\sin^2\theta_C$ $f_3^{\bar{\nu}p}(x) - f_3^{\nu p} = -6(f^{ep}(x) - f^{en}(x))$ works for every x as a total cross section. We now see that the objects produced at the right (in the hadron fragmentation region) are the same on both sides for the proton but not for the neutron - for the en experiment we must observe the isospin reflected products. If this is done, the relation holds for the partial cross sections for any products to the right if the left products are not observed.

Finally to bring all the hypotheses we have made about partons together into one list we note finally our suggestion that the f^{en}/f^{ep} rates strongly suggests that when a proton has a quark near x = 1 and a remainder of small momentum, that quark is a u quark. We try to generalize this to any baryon of the 56 multiplet in SU_6 language meaning our SU_6 only qualitatively, not exactly quantitatively.

E5. The amplitude that a baryon of the fundamental 56 is a parton of x near 1 and a remainder of small x varies as $(1-x)^{\gamma}$, the lowest γ occurs in the case that the parton is a quark, hence ③ in SU_3 (not an antiquark) and the remainder is $\bar{③}$, the ③ and $\bar{③}$ making the octet.

A member of the decimet Δ, has a smaller probability of having a quark near $x = 1$ than does the octet. We have already discussed the implications for total cross sections. It has implications for products also, of course.

We also assume that if a state is a pure quark to the left it has an amplitude to be a baryon to the left with z nearly 1 which is proportional the the chance the baryon contains a fast quark near $x = 1$ of the same kind. There are then many implications for products - and we have mentioned some that come from $u(x)$ being larger than all the others as $x \to 1$. There are others, of another type, for example (Cisneros, private communication) in the $e^+e^- \to$ hadrons since $u\bar{u}$ has four times the probability of $d\bar{d}$ the chance of producing a proton with x near 1(in the center of mass) and anything else is four times the probability of producing a neutron.

If E5 is correct we should like to assume something analogous for the mesons. The analogous assumption is that when one quark takes most of the momentum it is of a type that the low energy quark model supposes the meson to be made of. We therefore (in agreement with Cisneros) assume:

E6. For mesons with one parton near $x = 1$ it is a quark ③ and the remainder is $\bar{③}$ with spin 1/2, plus the charge conjugate with the same probability (for the charge conjugate it is $\bar{③}$ near $x = 1$ and the remainder is ③ .

We have built a very tall house of cards making so many weakly-based conjectures one upon the other and a great deal may be wrong. (Probably the weakest is C6 - same plateau for gap and hadron - but if it were wrong it does not alter the thrust of any of the others - just in the operator expressions we shall have to be careful to use the right plateaus.) Nevertheless this is the best guess I can make now - and we can try to use them has working hypotheses.

Probably the greatest challenge to experiment and theory is to get some evidence of quark quantum numbers in high-energy collisions. The low-energy quark model, good as it is, is not enough, there is always lingering doubt that the regularities observed have some entirely different basis or are, in part, accidental. The establishment of evidence for the quark model (and we have indicated very many ways - both in the last few lectures as well as earlier -(Llewellyn Smith's sum rule (Equation 33.6), the spin sum rule for $g_{1p}-g_{1n}$ etc.) by high energy experiments would confirm at once the reality of the regularities interpreted by the low energy quark model. This would make firm a conjecture of deepest significance to understanding high energy physics - the importance of quark quantum numbers.

Supposing for a moment this is done, then the next serious question will become theoretical - what exactly is the relation of the quark qualities at high energy and those at low energy. The "partons as quarks" model does not imply the low energy model (i.e. why are the wave functions not more complicated involving quark antiquark pairs) nor vice versa. At present their relation would not be understood. To start working on this now will take a little courage - you might waste your time - maybe partons as quarks will not be confirmed. If you do start, possibly one place to start might be to think about low energy matrix elements like $\Delta \to p + \gamma$ in a fast moving system in which all (or some) momenta are of the order P so parton wave functions can be used. (We have one relation of this kind in Bjorken's sum rule for g, Equation 33.16.)

Finally it should be noted that even if our house of cards survives and proves to be right, we have not thereby proved the existence of partons. The final result of our considerations has been to describe the result of the operation of a current on a proton state $J_\mu|p\rangle$ (for large ν, $-q^2 = 2M\nu x$) as a linear combination of operators like $D^L_\alpha E^R_{(p-\alpha,x)}{}^M|VAC$, creating final outgoing hadron states only. It might be wise to follow this out formally without mentioning partons (analogous to the way Gell-Mann and Fritzsch describe parton results for total inclusive cross sections in terms of commutation rules for quantities, currents, defined in general whether partons "exist" or not).

From this point of view the partons would appear as an unnecessary scaffolding that was used in building our house of cards.

On the other hand, the partons would have been a useful psychological guide as to what relations to expect - and if they continued to serve this way to produce other valid expectations they would of course begin to become "real", possibly as real as any other theoretical structure invented to describe Nature.

At any rate we shall see. It is good to have something to look forward to.

Appendix A. The Isospin of Quark Fragmentatior Products

APPENDIX A

The Isospin of Quark Fragmentation Products

The discussion (Lecture 56) leading to the idea that additive quark quantum numbers could appear as mean total quantum numbers of products moving in one direction is surprising - especially when it is noted that what holds for 3-isospin holds also for any other component such as 1-isospin or 2-isospin (although of course in practice they are nearly impossible to measure). It looks like an isospin 1/2 object could be represented by a group of isospin 1 objects (e.g. pions) - which at first seems impossible - except that we have an indefinite number of such objects.

It is therefore of interest to make a very simple special mathematical model, to show that indeed such things can be done in principle.

This is especially important when it is realized that our previous attempts at mathematical formulation cannot be complete and must be looked at as mere mnemonics (see note in Lecture 55 on the D and E operators). This little model may help by its example to lead to correct possible formal expressions of our ideas.

In this model suppose quarks carry only isospin 1/2 and hadrons are only pions of isospin 1 - made of quark antiquark pairs.

Imagine that we start with some current annihilation (analogous to e^+e^- but in more general isospin direction) initially disintegrating into a pair of quarks Q_α, $\bar{Q}_\beta$ (α,β are SU_2 indices fixed in this problem. Immediately after interaction:

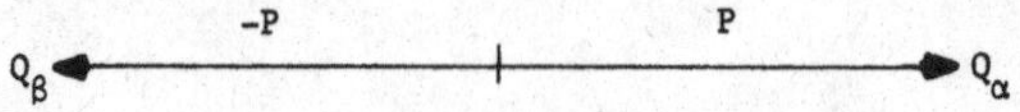

after Hamiltonian operates: N quark pairs in singlet state

$$\underbrace{Q_\beta \quad Q_i}\ \underbrace{\bar{Q}_i \quad Q_j}\ \bar{Q}_j \ldots Q\bar{Q} \ldots Q\ \underbrace{\bar{Q} \quad Q_\alpha}$$

makes hadrons (pions)	π_1	π_2	π_N
of type	$\vec{V}_1$	$\vec{V}_2$	$\vec{V}_N$

Next the action of the Hamiltonian is to produce pairs of quarks in a singlet state uniformly spaced in y space - a typical one is $Q_i\bar{Q}_i$ summed on all i equally. The number of such pairs N is then proportional to $\ell n 2P$ which we take to be very large. (One could also assume the number distributed via Poisson with a mean N etc., but we avoid complications which only serve to confuse our point - choose N fixed.)

Next this row of quarks is assumed to convert to pions by a simple rule, each pion is formed by a pair adjacent in the y space. Thus (in figure) if the first new singlet pair had index i, the next j etc. the first π is formed from an antiquark index β and a quark, index i; - the next by an antiquark index i and a quark index j; - etc.

To describe the isospin state of a π we use an isospin 3-vector $\vec{V}$. Thus if the π is a neutral pion π^0 we have $\vec{V}$ with only a z component, $\vec{V} = (0,0,1)$. For a π^+ we have $\vec{V} = \frac{1}{\sqrt{2}}(1,i,0)$ etc. The amplitude that an antiquark of index γ and a quark of index δ form a π characterized by vector $\vec{V}$ is then proportional to the $\gamma\delta$ matrix element of the two-by-two matrix $\sigma\cdot\vec{V}$ where σ are the Pauli matrices. Write this as $\langle\gamma|\sigma\cdot\vec{V}|\delta\rangle$. (We work in relative amplitudes and probabilities leaving overall normalization to the end.)

Thus the total amplitude to find the π's in directions $\vec{V}_1\vec{V}_2\ldots\vec{V}_N$ is

$$\text{Amp} = \sum_{i,j\ldots} \langle\beta|\sigma\cdot\vec{V}_1|i\rangle\ \langle i|\sigma\cdot\vec{V}_2|j\rangle\ \langle j|\sigma\cdot\vec{V}_3 \ \ldots\ldots\ \sigma\cdot\vec{V}_N|\alpha\rangle$$, the sums on

i, j etc., being because the newly found $\bar{Q}Q$ pairs are in singlet states. This is of course

$$\mathrm{Amp}(\vec{V}_1\ldots\vec{V}_N) = \langle\beta|(\sigma\cdot\vec{V}_1)(\sigma\cdot\vec{V}_2)\ldots(\sigma\cdot\vec{V}_N)|\alpha\rangle \tag{A1}$$

Having the amplitude (in an SU_2 invariant form, of course) we can ask many questions. The relative probability of finding any configuration is the square

$$P(\vec{V}_1\ldots\vec{V}_N) = \mathrm{Tr}\left[\rho_\beta(\sigma\cdot\vec{V}_1)(\sigma\cdot\vec{V}_2)\ldots(\sigma\cdot\vec{V}_N)\rho_\alpha(\sigma\cdot\vec{V}_N^*)\ldots(\sigma\cdot\vec{V}_2^*)(\sigma\cdot\vec{V}_1^*)\right] \tag{A2}$$

where ρ_α, ρ_β are 2x2 density matrices corresponding to the states, say $\rho_\alpha = a+\sigma\cdot\vec{A}$, $\rho_\beta = b+\sigma\cdot\vec{B}$. For state $|\alpha\rangle$, $\rho_\alpha = |\alpha\rangle\langle\alpha|$; for example if α is 1/2 in z, $|\alpha\rangle = \begin{pmatrix}1\\0\end{pmatrix}$ then $\rho_\alpha = \begin{pmatrix}1&0\\0&0\end{pmatrix} = \frac{1}{2}+\frac{1}{2}\sigma_z$.

Thus in that case $a = 1/2$, $A_z = 1/2$, $A_x = 0$, $A_y = 0$. If α is a state spinning in the direction of a unit vector then $\vec{A}/a$ is that unit vector. The expected value of z isospin is evidently $A_z/2a$.

Now suppose we observe the isospin character of only a limited number of pions, summing over the character of the rest. In fact we shall do two cases; summing over all to get the normalization of our probability, and summing over all but one, as if we studied products π number k + anything. In any case a sum over an unobserved π means a sum on V over 3 perpendicular values, symbolized by $\sum_V$ $\left(\text{i.e. } \sum_V (A\cdot V)(B\cdot V) = A\cdot B\right)$. We need the formula

$$\sum_V (\sigma\cdot\vec{V}^*)(a+\sigma\cdot\vec{B})(\sigma\cdot\vec{V}) = 3a - \sigma\cdot\vec{B} \tag{A3}$$

which is easily verified.

Now find the normalization $\mathcal{N} = \sum_{V_1V_2\ldots V_N} P(V_1\ldots V_N)$. On summing over V_1, ρ_β is converted from $b+\sigma\cdot\vec{B}$ to $3b-\sigma\cdot\vec{B}$ by (A3). Next sum V_2 and it is converted to $3^2b+\sigma\cdot\vec{B}$. Continue for N terms to $3^Nb+(-1)^N\sigma\cdot B$ whose trace with ρ_α gives (suppose tr(1) = 1)

$$\mathcal{N} = 3^N ab + (-1)^N(\vec{A}\cdot\vec{B}) \tag{A4}$$

For large N this is almost exactly 3^Nab and so we shall divide by this to get normalized probabilities. Thus the normalized probability that the kth pion is of type V_k is

$$P_k(V_k) = \frac{1}{3^N ab} \sum_{\substack{V_1 \ldots V_N \\ \text{except } V_k}} \mathrm{Tr}\Big[\rho_\beta(\sigma\cdot\vec{V}_1)\ldots(\sigma\cdot\vec{V}_k)\ldots(\sigma\cdot\vec{V}_N)\rho_\alpha(\sigma\cdot\vec{V}_N{}^*)\ldots$$

$$(\sigma\cdot\vec{V}_k{}^*)\ldots(\sigma\cdot\vec{V}_1)\Big]$$

Now again we can sum over $\vec{V}_1$, $\vec{V}_2$ up to $\vec{V}_{k-1}$ converting ρ_β to $3^{k-1}b+(-1)^{k-1}$ $(\sigma\cdot\vec{B})$; and also sum in an analogous way first on $\vec{V}_N$, then $\vec{V}_{N-1}$ etc. to $\vec{V}_{k+1}$ to convert ρ_α to $3^{N-k}a+(-1)^{N-k}(\sigma\cdot\vec{A})$. Thus the net is (scale tr(1) = 1)

$$P_k(V_k) = \frac{1}{3}\,\mathrm{Tr}\left[\left(1+\left(-\frac{1}{3}\right)^{k-1}\frac{\sigma\cdot\vec{B}}{b}\right)\left(\sigma\cdot\vec{V}_k\right)\left(1+\left(-\frac{1}{3}\right)^{N-k}\frac{\sigma\cdot\vec{A}}{a}\right)\left(\sigma\cdot\vec{V}_k{}^*\right)\right] \tag{A5}$$

from which any question about the one particle distribution function can now be answered. For example, sum over all possibilities for $\vec{V}_k$ and check normalization (for large N).

The mean probability the kth meson is π^+ is $P_k^{\pi^+}$ obtained by setting $\sigma\cdot\vec{V} = \frac{1}{\sqrt{2}}\,(\sigma_x+i\sigma_y)$ in (A5). For π^- put $\sigma\cdot\vec{V} = \frac{1}{\sqrt{2}}(\sigma_x-i\sigma_y)$. Thus the mean isospin of the kth hadron is $P_k^{\pi^+} - P_k^{\pi^-}$ = a term like (A5) with $\sigma\cdot\vec{V}_k\ldots\sigma\cdot\vec{V}_k{}^*$ replaced by $-i\sigma_x\ldots\sigma_y$ plus $i\sigma_y\ldots\sigma_x$. Note first using only the one coming from ρ_α this is $2\sigma_z$ and hence gives $\frac{2}{3}\left(-\frac{1}{3}\right)^{k-1} B_z/b$, likewise the term in $\sigma\cdot\vec{A}$ only contributes if σ is in the z direction.

We find

$$P_k^{\pi^+} - P_k^{\pi^-} = \frac{2}{3}\left(-\frac{1}{3}\right)^{k-1}\frac{B_z}{b} + \frac{2}{3}\left(-\frac{1}{3}\right)^{N-k}\frac{A_z}{a} \tag{A6}$$

This result confirms all our expectations. First if k is in the middle of the plateau, of order N/2, not near either end, then $P^{\pi^+} - P^{\pi^-} \sim \frac{1}{3}^{N/2}$ which is very small. Thus the plateau region has become neutral. $P_k^{\pi^+} - P_k^{\pi^-}$ can only avoid being small if either k is small (i.e. near the β end) or near N (i.e. near the α end). In the former case neglecting terms of order 3^{-N} we have

$$P_k^{\pi^+} - P_k^{\pi^-} = \frac{2}{3}\left(-\frac{1}{3}\right)^{k-1}\frac{B_z}{b} \tag{A7}$$

if k is finite, near the β end, a result that depends only on β, the quark at that end, and practically (as $N \to \infty$) not at all on the quark at the other end. (Evidently for k near N we have the exact opposite result.)

Finally the total z-isospin quantum number of all the pions to the left is

$$\sum_k (P_k^{\pi^+} - P_k^{\pi^-}) = \sum_{k=1} \frac{2}{3}\left(-\frac{1}{3}\right)^{k-1} \frac{B_z}{b} = \frac{1}{2}\frac{B_z}{b}$$

the z-component quantum number of the left-moving quark! The sum on k is to be taken from k = 1 to near N/2 to hold only the left movers, the contribution from the α-dependent term as only of order $3^{-N/2}$, and the sum is of the same order; the same as if we summed the left term above to infinity. Obviously the result is insensitive to exactly where in the plateau we stop, it is only necessary that we stop the sum on k at some point far from either end.

(One might further notice that $P_k^{\pi^+} + P_k^{\pi^-} - 2P_k^{\pi^0}$ is zero, neglecting 3^{-N}, for every k. This is a consequence of the fact that the quark has total isospin 1/2, and not higher. We leave to the reader to show that this generalizes to the real case and could in principle become a test of the total isospin character of the partons. That is $N^{\pi^+}_{(z)} + N^{\pi^-}_{(x)} - 2N^{\pi^o}_{(z)}$ is zero for any z to the left if the left movers come from a single parton (of any kind or superposition) and if the isospin of partons is either zero or one-half.)

Appendix B. A Test of Partons as Quarks

APPENDIX B

A Test of Partons as Quarks

Following J. D. Bjorken we note that the sum of neutrino and anti-neutrino cross-sections can be predicted fairly closely in terms of quantities already known. Since measurements of these total cross sections are the easiest test of quark quantum numbers, we give an analysis here.

Let us measure all cross sections in units of $Gs/2\pi$ where G is the Fermi constant and s the square of the center of mass energy. For nucleons, then, our unit is GEM/π where E is the laboratory energy.

The total cross section of a neutrino with a spin 1/2 particle is 2. With an antiparticle it is 2/3. Hence on a proton the cross section is

$$\sigma^{\nu p} = 2\int_0^1 x(d + \tfrac{1}{3}\,\bar{u})\,dx \quad ,$$

the factor x coming because the cross section varies with s. For neutrons we replace d by u, etc., so the mean neutrino cross section of a nucleon is

$$\sigma = \frac{1}{2}(\sigma^{\nu p} + \sigma^{\nu n}) = \int_0^1 x(d + u + \frac{1}{3}(\bar{d} + \bar{u}))dx.$$

the anti-neutrino cross section is

$$\bar{\sigma} = \frac{1}{2}(\sigma^{\bar{\nu}p} + \sigma^{\bar{\nu}n}) = \int_0^1 x(\bar{d} + \bar{u} + \frac{1}{3}(d + u))dx$$

Since $\bar{d}$, $\bar{u}$ are positive, but undoubtedly less than d, u, we see that $\bar{\sigma}/\sigma$ must be substantially less than 1 but greater than 1/3.

The sum is

$$\sigma + \bar{\sigma} = \frac{4}{3}\int_0^1 x(u + \bar{u} + d + \bar{d})dx$$

However, integrating the sum of (31.3) and (31.4) we have

$$\int_0^1 x(f^{ep} + f^{en})dx = \frac{5}{9}\int_0^1 x(u + \bar{u} + d + \bar{d})dx + \frac{2}{9}\int_0^1 x(s + \bar{s})dx$$

Experimentally this integral is 0.31 so if we could forget the integral $\int x(s+\bar{s})dx$ we would have $\sigma+\bar{\sigma} = \frac{4}{3}\cdot\frac{9}{5}(.31) = 0.74$. But $s+\bar{s}$ must surely be less than $u+\bar{u}$ and $d+\bar{d}$ and, when weighed by x, surely much less. It would be hard to manage to make inclusion of the last term produce more than a 10% effect. Thus we have a very stringent test of our parton quark model: $\sigma+\bar{\sigma}$ cannot exceed 0.74 and yet almost surely cannot fall below 0.74 by more than 10%.

One can also calculate upper limits for $\sigma^{\nu p} + \sigma^{\bar{\nu}p}$ and $\sigma^{\nu n} + \sigma^{\bar{\nu}n}$ separately (using other proportions of f^{ep} and f^{en}); they are 0.64 and 0.84 respectively.

These numerical estimates must be revised by a few percent, for we have neglected $\sin^2\theta_c$. They are valid only at asymptotic energy, of course, but T.D. Lee has pointed out that electron data indicates that this should only require a few GeV.

INDEX